Letts
EDUCATIONAL

ADVANCED
SUBSIDIARY

Revise AS
Mathematics

Author

Peter Sherran

Contents

Specification lists   4

AS/A2 Level Mathematics courses   9

Exam technique   12

What grade do you want?   15

Four steps to successful revision   17

Chapter 1 Pure 1

1.1 Algebra   19

1.2 Coordinate geometry   30

1.3 Sequences and series   33

1.4 Trigonometry   35

1.5 Differentiation   38

1.6 Integration   41

Sample questions and model answers   43

Practice examination questions   45

Chapter 2 Pure 2

2.1 Algebra and functions   48

2.2 Trigonometry   56

2.3 Differentiation   60

2.4 Integration   62

2.5 Numerical methods   64

Sample questions and model answers   67

Practice examination questions   69

Chapter 3 Mechanics 1

3.1	Vectors	72
3.2	Kinematics	75
3.3	Statics	79
3.4	Dynamics	84
	Sample questions and model answers	88
	Practice examination questions	90

Chapter 4 Statistics 1

4.1	Representing data	94
4.2	Probability	96
4.3	Discrete random variables	99
4.4	The Normal distribution	102
4.5	Correlation and regression	104
	Sample questions and model answers	106
	Practice examination questions	108

Chapter 5 Decision Mathematics 1

5.1	Algorithms	111
5.2	Graphs and networks	114
5.3	Critical path analysis	120
5.4	Linear programming	122
5.5	Matchings	125
	Sample questions and model answers	126
	Practice examination questions	128
	Practice examination answers	131
	Index	143

Specification lists

AQA Mathematics

MODULE	SPECIFICATION TOPIC	CHAPTER REFERENCE	STUDIED IN CLASS	REVISED	PRACTICE QUESTIONS
Pure 1	Algebra and functions	1.1			
	Coordinate geometry	1.2			
	Sequences and series	1.3			
	Trigonometry	1.4			
	Differentiation	1.5			
	Integration	1.6			
	Numerical methods	2.5			
Pure 2	Algebra and functions	2.1			
	Sequences and series	1.3, 2.1			
	Trigonometry	2.2			
	Exponentials and logarithms	2.1			
	Differentiation	2.3			
	Integration	2.4			
Mechanics 1	Application of vectors	3.1			
	Forces	3.3			
	Kinematics	3.2			
	Newton's laws	3.4			
	Connected particles	3.4			
	Moments and centres of mass	3.3			
	Projectiles	3.2			
	Momentum	3.4			
Statistics 1	Sampling	4.1			
	Probability	4.2			
	Descriptive statistics	4.1			
	Discrete probability distributions	4.3			
	Normal distribution	4.4			
	Correlation and regression	4.5			
Decisions 1	Graphs	5.2			
	Networks	5.2			
	Critical path analysis	5.3			
	Linear programming	5.4			
	Boolean algebra				

Examination analysis

The *assessment is by written papers. All questions are compulsory.*
Pure 1 + Pure 2 + one of Mechanics 1, Statistics 1 and Decision 1.

Pure 1	AS	May use graphic calculator	1 hr 15 min exam	30%
Pure 2	AS	Scientific calculator only	1 hr 15 min exam	30%
Mechanics 1	AS	May use graphic calculator	1 hr 45 min exam	40%
Statistics 1	AS	May use graphic calculator	1 hr 45 min exam	40%
Decision 1	AS	May use graphic calculator	1 hr 45 min exam	40%

Edexcel Mathematics

MODULE	SPECIFICATION TOPIC	CHAPTER REFERENCE	STUDIED IN CLASS	REVISED	PRACTICE QUESTIONS
Pure 1	Algebra	1.1			
	Trigonometry	1.4			
	Coordinate geometry	1.2			
	Sequences and series	1.3			
	Differentiation	1.5			
	Integration	1.6			
Pure 2	Algebra and functions	2.1			
	Functions	2.1			
	Sequences and series	1.3, 2.1			
	Trigonometry	2.2			
	Exponentials and logarithms	2.1			
	Differentiation	2.3			
	Integration	2.4			
	Numerical methods	2.5			
Mechanics 1	Vectors in mechanics	3.1			
	Kinematics in a straight line	3.2			
	Dynamics in a straight line or plane	3.4			
	Statics of a particle	3.3			
	Moments	3.3			
Statistics 1	Representation and summary of data	4.1			
	Probability	4.2			
	Correlation and regression	4.5			
	Discrete random variables	4.3			
	Discrete distributions	4.3			
	The Normal distribution	4.4			
Decisions 1	Algorithms	5.1			
	Algorithms on graphs	5.2			
	The route inspection problem	5.2			
	Critical path analysis	5.3			
	Linear programming	5.4			
	Matchings	5.5			
	Flows in networks	5.2			

Examination analysis

The *assessment is by written papers. All questions are compulsory.*
Pure 1 + Pure 2 + one of Mechanics 1, Statistics 1 and Decision 1.

Pure 1	AS	Scientific calculator only	1 hr 30 min exam	33.3%
Pure 2	A2	May use graphic calculator	1 hr 30 min exam	33.3%
Mechanics 1	AS	May use graphic calculator	1 hr 30 min exam	33.3%
Statistics 1	AS	May use graphic calculator	1 hr 30 min exam	33.3%
Decision 1	AS	May use graphic calculator	1 hr 30 min exam	33.3%

OCR Mathematics

MODULE	SPECIFICATION TOPIC	CHAPTER REFERENCE	STUDIED IN CLASS	REVISED	PRACTICE QUESTIONS
Pure 1	Indices and surds	1.1			
	Quadratics	1.1			
	Coordinate geometry and graphs	1.2			
	Trigonometry	1.4			
	Differentiation	1.5			
	Integration	1.6			
Pure 2	Circular measure	1.4			
	Sequences and series	1.3, 2.1			
	Polynomials	2.1			
	Functions, the modulus function	2.1			
	Logarithmic and exponential functions	2.1			
	Differentiation and integration	2.3, 2.4			
	Numerical methods	2.5			
Mechanics 1	Force as a vector	3.1			
	Equilibrium of a particle	3.3			
	Kinematics of motion in a straight line	3.2			
	Newton's laws of motion	3.4			
	Linear momentum	3.4			
Statistics 1	Representation of data	4.1			
	Probability	4.2			
	Discrete random variables	4.3			
	Bivariate data	4.5			
Decisions 1	Algorithms	5.1			
	Graph theory	5.2			
	Networks	5.2			
	Linear programming	5.4			

Examination analysis

The *assessment is by written papers. All questions are compulsory.*
Pure 1 + Pure 2 + one of Mechanics 1, Statistics 1 and Decision 1.

Pure 1	AS	Scientific calculator only	1 hr 20 min exam	33.3%
Pure 2	A2	May use graphic calculator	1 hr 20 min exam	33.3%
Mechanics 1	AS	May use graphic calculator	1 hr 20 min exam	33.3%
Statistics 1	AS	May use graphic calculator	1 hr 20 min exam	33.3%
Decision 1	AS	May use graphic calculator	1 hr 20 min exam	33.3%

WJEC Mathematics

MODULE	SPECIFICATION TOPIC	CHAPTER REFERENCE	STUDIED IN CLASS	REVISED	PRACTICE QUESTIONS
Pure 1	Algebra	1.1			
	Coordinate geometry and graphs	1.2			
	Trigonometry	1.4			
	Differentiation	1.5			
	Integration	1.6			
Pure 2	Polynomials	1.1			
	Functions	2.1			
	Logarithmic and exponential functions	2.1			
	Differentiation	2.3			
	Integration	2.4			
	Numerical methods	2.5			
Mechanics 1	Motion with uniform acceleration	3.2			
	Vertical motion under gravity	3.2			
	Dynamics of a particle	3.4			
	Statics	3.3			
Statistics 1	Representation of data	4.1			
	Probability	4.2			
	Discrete probability distributions	4.3			
	Binomial and Poisson distributions	4.3			

Examination analysis

The *assessment is by written papers. All questions are compulsory.*
Pure 1 + Pure 2 + one of Mechanics 1, Statistics 1 and Decision 1.

Pure 1	AS	Scientific calculator only	1 hr 30 min exam	33.3%
Pure 2	A2	Scientific calculator only	1 hr 30 min exam	33.3%
Mechanics 1	AS	May use graphic calculator	1 hr 30 min exam	33.3%
Statistics 1	AS	May use graphic calculator	1 hr 30 min exam	33.3%

NICCEA Mathematics

MODULE	SPECIFICATION TOPIC	CHAPTER REFERENCE	STUDIED IN CLASS	REVISED	PRACTICE QUESTIONS
Pure 1	Algebra and functions	1.1			
	Sequences and series	1.3			
	Trigonometry	1.4			
	Calculus	1.5, 1.6			
Pure 2	Algebra and functions	2.1			
	Coordinate geometry in the (x, y) plane	1.2			
	Sequences and series	1.3, 2.1			
	Trigonometry	2.2			
	Exponentials and logarithms	2.1			
	Calculus	2.3, 2.4			
	Numerical methods	2.5			
Mechanics 1	Kinematics of a particle (straight line)	3.2			
	Statics	3.3			
	Dynamics	3.4			
Statistics 1	Mathematics of uncertainty	4.2			
	Discrete probability distributions	4.3			
	Continuous probability distributions	4.4			

Examination analysis

The *assessment is by written papers. All questions are compulsory.*
Pure 1 + Pure 2 + one of Mechanics 1, Statistics 1 and Decision 1.

Pure 1	AS	Scientific calculator only	1 hr 30 min exam	33.3%
Pure 2	A2	Scientific calculator only	1 hr 30 min exam	33.3%
Mechanics 1	AS	May use graphic calculator	1 hr 30 min exam	33.3%
Statistics 1	AS	May use graphic calculator	1 hr 30 min exam	33.3%

AS/A2 Level Mathematics courses

AS and A2

All Mathematics A Level courses being studied from September 2000 are in two parts, with three separate modules in each part. Students first study the AS (Advanced Subsidiary) course. Some will then go on to study the second part of the A Level course, called A2. Advanced Subsidiary is assessed at the standard expected halfway through an A Level course: i.e., between GCSE and Advanced GCE. This means that new AS and A2 courses are designed so that difficulty steadily increases:

- AS Mathematics builds from GCSE Mathematics
- A2 Mathematics builds from AS Mathematics.

How will you be tested?

Assessment units

For AS Mathematics, you will be tested by three assessment units. For the full A Level in Mathematics, you will take a further three units. AS Mathematics forms 50% of the assessment weighting for the full A Level.

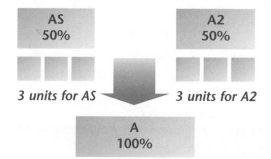

Each unit can normally be taken in either January or June. Alternatively, you can study the whole course before taking any of the unit tests. There is a lot of flexibility about when exams can be taken and the diagram below shows just some of the ways that the assessment units may be taken for AS and A Level Mathematics.

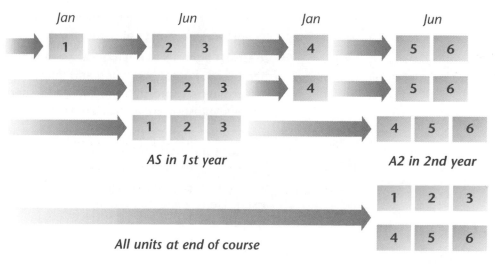

If you are disappointed with a module result, you can resit each module once. You will need to be very careful about when you take up a resit opportunity because you will have only one chance to improve your mark. The higher mark counts.

A2 and Synoptic assessment

After having studied AS Mathematics, you may wish to continue studying Mathematics to A Level. For this you will need to take three further units of Mathematics at A2. Similar assessment arrangements apply except some units, those that draw together different parts of the course in a 'synoptic' assessment, have to be assessed at the end of the course.

Coursework

Coursework may form part of your A Level Mathematics course, depending on which specification you study. Where students have to undertake coursework, it is usually for the assessment of practical skills but this is not always the case.

Key skills

It is important that you develop your key skills throughout your AS and A2 courses. These are important skills that you need whatever you do beyond AS and A Levels. To gain the key skills qualification, which is equivalent to an AS Level, you will need to collect evidence together in a 'portfolio' to show that you have attained Level 3 in Communication, Application of number and Information technology. You will also need to take a formal testing in each key skill. You will have many opportunities during AS Mathematics to develop your key skills.

It is a worthwhile qualification, as it demonstrates your ability to put your ideas across to other people, collect data and use up-to-date technology in your work.

What skills will I need?

For AS Mathematics, you will be tested by assessment objectives: these are the skills and abilities that you should have acquired by studying the course. The assessment objectives for AS Mathematics are shown below.

Candidates should be able to:

- recall, select and use their knowledge of mathematical facts, concepts and techniques in a variety of contexts

- construct rigorous mathematical arguments and proofs through use of precise statements, logical deduction and inference and by the manipulation of mathematical expressions, including the construction of extended arguments for handling substantial problems presented in unstructured form

- recall, select and use their knowledge of standard mathematical models to represent situations in the real world; recognise and understand given representations involving standard models; present and interpret results from such models in terms of the original situation, including discussion of the assumptions made and refinement of such models

- comprehend translations of common realistic contexts into Mathematics; use the results of calculations to make predictions, or comment on the context; and, where appropriate, read critically and comprehend longer mathematical arguments or examples of applications

- use contemporary calculator technology and other permitted resources (such as formulae booklets or statistical tables) accurately and efficiently; understand when not to use such technology, and its limitations; give answers to appropriate accuracy.

Exam technique

Mathematics is, inherently, a sequential subject. There is a progression of material through all levels at which the subject is studied. The criteria therefore build on the knowledge, understanding and skills established at GCSE.

Thus, candidates embarking on AS/A2 Level study in Mathematics subjects are expected to have achieved at least Grade C in GCSE Mathematics, or equivalent, and to have covered all the material in the Intermediate Tier. In addition, candidates will be expected to be able to use the material listed below whenever it is required. This material, together with the GCSE material, is regarded as assumed background knowledge. However, it may be assessed within questions focused on other material from the relevant specification.

Background knowledge

a The arithmetic of integers (including HCFs and LCMS), of fractions, and real numbers.

b The laws of indices for positive integer exponents.

c Solution of problems involving ratio and proportion (including similar triangles, and links between length, area and volume of similar figures).

d Elementary algebra (including multiplying out brackets, factorising quadratics with integer coefficients, to include $a^2 - b^2$, and solution of simultaneous linear equations by eliminating a variable).

e Changing the subject of a simple formula or equation.

f The equation $y = mx + c$ for a straight line; gradient and intercept.

g The distance between two points in 2-D with given coordinates.

h Solution of triangles using trigonometry, including the sine and cosine rules.

i Volume of cone and sphere.

j The following properties of a circle:

   (i) the angle in a semicircle is a right angle
   (ii) the perpendicular from the centre to a chord bisects the chord
   (iii) the perpendicularity of radius and tangent.

What are the examiners looking for?

Examiners use certain words in their instructions to let you know what they are expecting in your answer. Make sure that you know what they mean so that you can give the right response.

Write down, state

You can write your answer without having to show how it was obtained. There is nothing to prevent you doing some working if it helps you, but if you are doing a lot then you might have missed the point.

Calculate, find, determine, show, solve

Make sure that you show enough working to justify the final answer or conclusion. Marks will be available for showing a correct method.

Deduce, hence

This means that you are expected to use the given result to establish something new. You must show all of the steps in your working.

Draw

This is used to tell you to plot an accurate graph using graph paper. Take note of any instructions about the scale that must be used. You may need to read values from your graph.

Sketch

If the instruction is to sketch a graph then you don't need to plot the points but you will be expected to show its general shape and its relationship with the axes. Indicate the positions of any turning points and take particular care with any asymptotes.

Find the exact value

This instruction is usually given when the final answer involves an irrational value such as a logarithm, e, π or a surd. You will need to demonstrate that you can manipulate these quantities so don't just key everything into your calculator or you will lose marks.

If a question requires the final answer to be given to a specific level of accuracy then make sure that you do this or you might needlessly lose marks.

Some dos and don'ts

Dos

Do read the question
- Make sure that you are clear about what you are expected to do. Look for some structure in the question that may help you take the right approach.
- Read the question *again* after you have answered it as a quick check that your answer is in the expected form.

Do use diagrams
- In some questions, particularly in mechanics, a clearly labelled diagram is essential. Use a diagram whenever it may help you understand or represent the problem that you are trying to solve.

Do take care with notation
- Write clearly and use the notation accurately. Use brackets when they are required.
- Even if your final answer is wrong, you may earn some marks for a correct expression in your working.

Do avoid silly answers

- Check that your final answer is sensible within the context of the question.

Do make good use of time

- Choose the order in which you answer the questions carefully. Do the ones that are easiest for you first.
- Set yourself a time limit for a question depending on the number of marks available.
- Be prepared to leave a difficult part of a question and return to it later if there is time.
- Towards the end of the exam make sure that you pick up all of the easy marks in any questions that you haven't got time to answer fully.

Don'ts

Don't work with rounded values

- There may be several stages in a solution that produce numerical values. Rounding errors from earlier stages may distort your final answer. One way to avoid this is to make use of your calculator memories to store values that you will need again.

Don't cross out work that may be partly correct

- It's tempting to cross out something that hasn't worked out as it should. Avoid this unless you have time to replace it with something better.

Don't write out the question

- This wastes time. The marks are for your solution!

What grade do you want?

Everyone should be able to improve their grades but you will only manage this with a lot of hard work and determination. The details given below describe a level of performance typical of candidates achieving grades A, C or E. You should find it useful to read and compare the expectations for the different levels and to give some thought to the areas where you need to improve most.

Grade A candidates

- Recall or recognise almost all the mathematical facts, concepts and techniques that are needed, and select appropriate ones to use in a variety on contexts.
- Manipulate mathematical expressions and use graphs, sketches and diagrams, all with high accuracy and skill.
- Use mathematical language correctly and proceed logically and rigorously through extended arguments or proofs.
- When confronted with unstructured problems they can often devise and implement an effective solution strategy.
- If errors are made in their calculations or logic, these are sometimes noticed and corrected.
- Recall or recognise almost all the standard models that are needed, and select appropriate ones to represent a wide variety of situations in the real world.
- Correctly refer results from calculations using the model to the original situation; they give sensible interpretations of their results in the context of the original realistic situation.
- Make intelligent comments on the modelling assumptions and possible refinements to the model.
- Comprehend or understand the meaning of almost all translations into mathematics of common realistic contexts.
- Correctly refer the results of calculations back to given context and usually make sensible comments or predictions.
- Can distil the essential mathematical information from extended pieces of prose having mathematical content.
- Comment meaningfully on the mathematical information.
- Make appropriate and efficient use of contemporary calculator technology and other permitted resources, and are aware of any limitations to their use.
- Present results to an appropriate degree of accuracy.

Grade C candidates

- Recall or recognise most of the mathematical facts, concepts and techniques that are needed, and usually select appropriate ones to use in a variety of contexts.
- Manipulate mathematical expressions and use graphs, sketches and diagrams, all with a reasonable level of accuracy and skill.
- Use mathematical language with some skill and sometimes proceed logically through extended arguments or proofs.
- When confronted with unstructured problems they sometimes devise and implement an effective and efficient solution strategy.

- Occasionally notice and correct errors in their calculations.
- Recall or recognise most of the standard models that are needed and usually select appropriate ones to represent a variety of situations in the real world.
- Often correctly refer results from calculations using the model to the original situation, they sometimes give sensible interpretations of their results in context of the original realistic situation.
- Sometimes make intelligent comments on the modelling assumptions and possible refinements to the model.
- Comprehend or understand the meaning of most translations into mathematics of common realistic contexts.
- Often correctly refer the results of calculations back to the given context and sometimes make sensible comments or predictions.
- Distil much of the essential mathematical information from extended pieces of prose having mathematical content.
- Give some useful comments on this mathematical information.
- Usually make appropriate and effective use of contemporary calculator technology and other permitted resources, and are sometimes aware of any limitations to their use.
- Usually present results to an appropriate degree of accuracy.

Grade E candidates

- Recall or recognise some of the mathematical facts, concepts and techniques that are needed, and sometimes select appropriate ones to represent to use in some contexts.
- Manipulate mathematical expressions and use graphs, sketches and diagrams, all with some accuracy and skill.
- Sometimes use mathematical language correctly and occasionally proceed logically through extended arguments or proofs.
- Recall or recognise some of the standard models that are needed and sometimes select appropriate ones to represent a variety of situations in the real world.
- Sometimes correctly refer results from calculations using the model to the original situation; they try to interpret their results in the context of the original realistic situation.
- Sometimes comprehend or understand the meaning of translations in mathematics of common realistic contexts.
- Sometimes correctly refer the results of calculations back to the given context and attempt to give comments or predictions.
- Distil some of the essential mathematical information from extended pieces of prose having mathematical content; they attempt to comment on this mathematical information.
- Candidates often make appropriate and efficient use of contemporary calculator technology and other permitted resources.
- Often present results to an appropriate degree of accuracy.

The table below shows how your average mark is translated.

average	80%	70%	60%	50%	40%
grade	A	B	C	D	E

Four steps to successful revision

Step 1: Understand

- Study the topic to be learned slowly. Make sure you understand the logic or important concepts.
- Mark up the text if necessary – underline, highlight and make notes.
- Re-read each paragraph slowly.

GO TO STEP 2

Step 2: Summarise

- Now make your own revision note summary:
  What is the main idea, theme or concept to be learned?
  What are the main points? How does the logic develop?
  Ask questions: Why? How? What next?
- Use bullet points, mind maps, patterned notes.
- Link ideas with mnemonics, mind maps, crazy stories.
- Note the title and date of the revision notes
  (e.g. Mathematics: Trigonometry, 3rd March).
- Organise your notes carefully and keep them in a file.

This is now in **short-term memory**. You will forget 80% of it if you do not go to Step 3.
GO TO STEP 3, but first take a 10 minute break.

Step 3: Memorise

- Take 25 minute learning 'bites' with 5 minute breaks.
- After each 5 minute break test yourself:
  Cover the original revision note summary
  Write down the main points
  Speak out loud (record on tape)
  Tell someone else
  Repeat many times.

The material is well on its way to **long-term memory**.
You will forget 40% if you do not do step 4. **GO TO STEP 4**

Step 4: Track / Review

- Create a Revision Diary (one A4 page per day).
- Make a revision plan for the topic, e.g. 1 day later, 1 week later, 1 month later.
- Record your revision in your Revision Diary, e.g.
  Mathematics: Trigonometry, 3rd March 25 minutes
  Mathematics: Trigonometry, 5th March 15 minutes
  Mathematics: Trigonometry, 3rd April 15 minutes
  ... and then at monthly intervals.

Chapter 1

Pure 1

The following topics are covered in this chapter:

- Algebra
- Coordinate geometry
- Sequences and series

- Trigonometry
- Differentiation
- Integration

1.1 Algebra

After studying this section you should be able to:

- work with indices and surds
- use function notation and understand domain and range
- solve quadratic equations and sketch their graphs
- recognise and sketch graphs of a range of functions
- use transformations to sketch graphs of related functions
- solve simultaneous equations – one linear and one quadratic
- work with identities
- use the factor theorem
- solve linear and quadratic inequalities

LEARNING SUMMARY

Indices

AQA	P1
EDEXCEL	P1
OCR	P1
WJEC	P1
NICCEA	P1

Remember that, in general
$(a+b)^n \neq a^n + b^n$
and $(a-b)^n \neq a^n - b^n$.

You need to know all of the basic rules and be able to apply them.

$$a^m \times a^n = a^{m+n} \qquad a^m \div a^n = a^{m-n} \qquad (a^m)^n = a^{mn} \qquad a^{\frac{1}{n}} = \sqrt[n]{a}$$

$$a^{\frac{m}{n}} = (a^m)^{\frac{1}{n}} = (a^{\frac{1}{n}})^m \qquad a^{-n} = \frac{1}{a^n} \qquad (ab)^n = a^n b^n \qquad \left(\frac{a}{b}\right)^n = \frac{a^n}{b^n}$$

Some important special cases are: $\quad a^1 = a \qquad a^{\frac{1}{2}} = \sqrt{a}$

and $\quad a^0 = 1 \qquad a^{-1} = \frac{1}{a} \quad$ provided $a \neq 0$.

For example $\quad 9^{\frac{1}{2}} = \sqrt{9} = 3 \quad$ and $\quad 16^{\frac{3}{2}} = (16^{\frac{1}{2}})^3 = 4^3 = 64.$

Surds

AQA	P1
EDEXCEL	P1
OCR	P1
WJEC	P1
NICCEA	P1

An irrational number is a number which continues for ever after the decimal point without a repeating pattern.

A **surd** is the square root of a whole number that has an **irrational** value.

Some examples are $\sqrt{2}$, $\sqrt{3}$ and $\sqrt{10}$.

You can often simplify a surd using the fact that $\sqrt{ab} = \sqrt{a} \times \sqrt{b}$.

This simplification works because 4 is a square number.

For example $\sqrt{12} = \sqrt{4 \times 3} = \sqrt{4} \times \sqrt{3} = 2\sqrt{3}.$

Surds may be used to express some results in exact form such as $\sin 60° = \frac{\sqrt{3}}{2}$.

Functions

AQA	P1
EDEXCEL	P1
OCR	P2
WJEC	P2
NICCEA	P1

A **function** may be thought of as a rule which takes each member x of a set and assigns, or **maps**, it to some value y known as its **image**.

$$x \longrightarrow \boxed{\text{Function}} \longrightarrow y$$

> x maps to y
> y is the image of x.

> $f(x)$ is read as 'f of x'.

A letter such as f, g or h is often used to stand for a function. The function which squares a number and adds on 5, for example, can be written as $f(x) = x^2 + 5$. The same notation may also be used to show how a function affects particular values. For this function, $f(4) = 4^2 + 5 = 21$, $f(-10) = (-10)^2 + 5 = 105$ and so on.

An alternative notation for the same function is $f: x \mapsto x^2 + 5$.

> **KEY POINT**
> The set of values on which the function acts is called the **domain** and the corresponding set of image values is called the **range**.

The function $f(x) = x + 2$ with domain {3, 6, 10} is shown below.

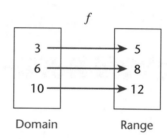

$$
\begin{array}{ccc}
 & f & \\
3 & \longrightarrow & 5 \\
6 & \longrightarrow & 8 \\
10 & \longrightarrow & 12 \\
\text{Domain} & & \text{Range}
\end{array}
$$

> The symbol $\mathbb{R}$ is often used to stand for the complete set of numbers on the number line. This is the set of Real numbers.

The domain of a function may be an infinite set such as $\mathbb{R}$ but, in some cases, particular values must be omitted for the function to be valid.

For example, the function $f(x) = \dfrac{x + 2}{x - 3}$ can not have 3 in its domain since division by zero is undefined.

Under a function, every member of the domain must have only one image. But, it is possible for different values in the domain to have the same image.

> There are different types of function.

> **KEY POINT**
> If for each element y in the range, there is *a unique* value of x such that $f(x) = y$ then f is a **one–one** function. If for any element y in the range, there is more than one value of x satisfying $f(x) = y$ then f is **many–one**.

Examples

> These examples show the importance of including the domain of the function in its definition.

The function $f(x) = 2x$ with domain $\mathbb{R}$ is one–one.

The function $f(x) = x^2$ with domain $\mathbb{R}$ is many–one.

The function $f(x) = x^2$ with domain $x > 0$ is one–one.

The function $f(x) = \sin x$ with domain $0° \leqslant x \leqslant 360°$ is many–one.

The function $f(x) = \sin x$ with domain $-90° \leqslant x \leqslant 90°$ is one–one.

Quadratics

AQA   P1
EDEXCEL   P1
OCR   P1
WJEC   P1
NICCEA   P1

In algebra, any expression of the form $ax^2 + bx + c$ where $a \neq 0$ is called a quadratic.

You need to know these basic results about quadratics.

$$(x + a)^2 = x^2 + 2ax + a^2 \qquad (x - a)^2 = x^2 - 2ax + a^2 \qquad (x + a)(x - a) = x^2 - a^2$$

Some examples are:

$$(x + 3)^2 = x^2 + 6x + 9 \qquad (x - 5)^2 = x^2 - 10x + 25$$
$$(x + 7)(x - 7) = x^2 - 49 \qquad (x + \sqrt{3})(x - \sqrt{3}) = x^2 - 3.$$

To solve problems in algebra you need to develop your skills so that you can recognise how to apply the basic results.

In this fraction, the denominator $2 - \sqrt{3}$ is irrational.

For example, the fraction $\dfrac{1}{2 - \sqrt{3}}$ may be simplified by **rationalising the denominator**. You can do this by multiplying the numerator and denominator by $2 + \sqrt{3}$ to make an equivalent fraction in which the denominator is rational.

$(2 - \sqrt{3})(2 + \sqrt{3})$
$= 2^2 - (\sqrt{3})^2$
$= 4 - 3.$

$$\frac{1}{2 - \sqrt{3}} = \frac{1}{2 - \sqrt{3}} \times \frac{2 + \sqrt{3}}{2 + \sqrt{3}}$$

$$= \frac{2 + \sqrt{3}}{4 - 3} \qquad \text{(The denominator is now rational)}$$

$$= 2 + \sqrt{3}.$$

Quadratic equations

AQA   P1
EDEXCEL   P1
OCR   P1
WJEC   P1
NICCEA   P1

Equations of the form $ax^2 + bx + c = 0$ (where $a \neq 0$) are quadratic equations.

Some quadratic equations can be solved by **factorising**.

Example   Solve $x^2 - 3x - 10 = 0$.
$$(x - 5)(x + 2) = 0 \qquad \text{This means that either } x - 5 = 0$$
$$\text{or } x + 2 = 0.$$

so $x = 5$ or $x = -2$.

If the quadratic will not factorise then you can try **completing the square**.

$x^2 - 6x + 9 = (x - 3)^2$ so
$x^2 - 6x + 1 = (x - 3)^2 - 8.$

Example   Solve $x^2 - 6x + 1 = 0$.
$$(x - 3)^2 - 8 = 0 \qquad \text{(Now } x \text{ only appears once in the equation)}$$
$$(x - 3)^2 = 8$$
$$x - 3 = \pm\sqrt{8}$$

$\sqrt{8} = \sqrt{(4 \times 2)} = 2\sqrt{2}.$

$$x = 3 \pm 2\sqrt{2}.$$
$$\text{(The solutions are } 3 + 2\sqrt{2} \text{ and } 3 - 2\sqrt{2})$$

The method shown for completing the square can be adapted for the general form of a quadratic. This gives the usual formula for the solution of a quadratic equation.

Multiply both sides by $4a$.

$$ax^2 + bx + c = 0$$
$$4a^2x^2 + 4abx + 4ac = 0$$

Note: $4a^2x^2 + 4abx + b^2 = (2ax + b)^2$.

Add b^2 to complete the square and subtract it again to keep things the same.

$$4a^2x^2 + 4abx + b^2 + 4ac - b^2 = 0$$
$$(2ax + b)^2 + 4ac - b^2 = 0$$
$$(2ax + b)^2 = b^2 - 4ac$$

Now x only appears once in the equation.

Find the square root of both sides.

$$2ax + b = \pm\sqrt{b^2 - 4ac}$$

Rearrange to find x.

$$x = \frac{-b \pm \sqrt{b^2 - 4ac}}{2a}$$

This is the **quadratic formula**.

Example   Solve $3x^2 - 4x - 9 = 0$ giving your answers to two decimal places.

Comparing this equation with the general form gives $a = 3$, $b = -4$ and $c = -9$. Substitute this information into the formula:

$$x = \frac{4 \pm \sqrt{(-4)^2 - 4 \times 3 \times (-9)}}{6} = \frac{4 \pm \sqrt{124}}{6}$$

$x = 2.52$ or -1.19 (two decimal places).

In the formula, the value of $b^2 - 4ac$ is called the **discriminant**. This value can be used to give information about the solutions without having to solve the equation.

In the example above, $b^2 - 4ac = 124 > 0$ and two real roots were found.

$b^2 - 4ac > 0$ *two* real solutions (the solutions are often called **roots**)

$b^2 - 4ac = 0$ *one* real solution (often thought of as one **repeated root**)

$b^2 - 4ac < 0$ *no* real solutions.

The solutions of a quadratic equation correspond to where its graph crosses the x-axis. There are three possible situations depending on the value of the discriminant.

The diagrams correspond to the situation where $a > 0$. The same principle applies when $a < 0$ but the diagrams appear the other way up.

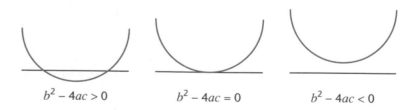

$b^2 - 4ac > 0$ $b^2 - 4ac = 0$ $b^2 - 4ac < 0$

Example   The equation $5x^2 + 3x + p = 0$ has a repeated root. Find the value of p.

In this case, $a = 5$, $b = 3$ and $c = p$.

For a repeated root $b^2 - 4ac = 0$ so $9 - 20p = 0$, giving $p = 0.45$.

Quadratic graphs

AQA — P1
EDEXCEL — P1
OCR — P1
WJEC — P1
NICCEA — P1

The graph of $y = ax^2 + bx + c$ is a parabola.

$a > 0$ $a < 0$

The methods used for solving quadratic equations can also be used to give information about the graphs.

Example   Sketch the graph of $y = x^2 - x - 6$.
Find the coordinates of the lowest point on the curve.

The curve will cross the x-axis when $y = 0$. You can find these points by solving the equation $x^2 - x - 6 = 0$.

> The curve will cross the y-axis when $x = 0$, giving $y = -6$.

$$x^2 - x - 6 = 0 \Rightarrow (x + 2)(x - 3) = 0$$

$$\Rightarrow x = -2 \quad \text{or} \quad x = 3.$$

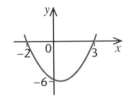

The curve is symmetrical so the lowest point occurs mid-way between -2 and 3 and this is given by $(-2 + 3) \div 2 = 0.5$.

When $x = 0.5$, $y = 0.5^2 - 0.5 - 6 = -6.25$.

> The vertex of the curve occurs at its maximum or minimum point.
>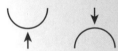

The lowest point on the curve is $(0.5, -6.25)$.

Completing the square gives information about the **vertex** of the curve even if the equation will not factorise.

Example   Find the co-ordinates of the vertex of the curve $y = x^2 + 2x + 3$.

You need to recognise that $x^2 + 2x + 1 = (x + 1)^2$,

then, completing the square, $x^2 + 2x + 3 = x^2 + 2x + 1 + 2 = (x + 1)^2 + 2$.

The equation of the curve can now be written as $y = (x + 1)^2 + 2$.

$(x + 1)^2$ cannot be negative so its minimum value is zero, when $x = -1$.

This means that the minimum value of y is 2 and this occurs when $x = -1$.

The vertex of the curve is at $(-1, 2)$.

More graphs

AQA   P1
EDEXCEL   P1
OCR   P1
WJEC   P1
NICCEA   P1

The graph of a cubic equation $y = ax^3 + bx^2 + cx + d$ can take a number of forms.

Use a graphic calculator to draw the graphs when:

$$b^2 > 3ac$$
$$b^2 = 3ac$$
$$b^2 < 3ac$$

and compare the results with these given forms.

$a > 0$ $a < 0$

The cubic curve has rotational symmetry of order 2 about the point where $x = -\dfrac{b}{3a}$.

> **KEY POINT**
>
> A cubic equation that can be factorised as $y = (x - p)(x - q)(x - r)$ will cross the x-axis at p, q and r. If any two of p, q and r are the same then the x-axis will be a tangent to the curve at that point.

For example, the graph of $y = (x + 2)(x - 3)^2$ looks like this:

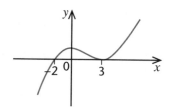

This shows a repeated root at $x = 3$.

> **KEY POINT**
>
> The graph of $y = x^n$ where n is an integer has:
> **rotational** symmetry about the **origin** when n is **odd**
> **reflective** symmetry in the y-axis when n is **even**.

The graphs of $y = x^3$, $y = x^5$, $y = x^7$ ... look something like this:

The graphs of $y - x^2$, $y - x^4$, $y = x^6$ ... look something like this:

For higher powers of x the graphs are flatter between -1 and 1 and steeper elsewhere.

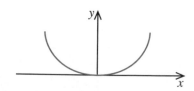

The graphs of $y = x^{-1}$, $y = x^{-3}$, $y = x^{-5}$ ... look something like this:

The graphs of $y = x^{-2}$, $y = x^{-4}$, $y = x^{-6}$ ... look something like this:

As the powers of x become more negative, the graphs become steeper between -1 and 1 and flatter elsewhere.

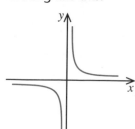

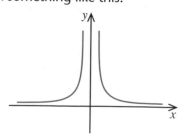

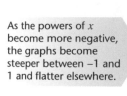

The **modulus function** $y = |ax + b|$ represents the positive value of the expression inside the brackets. The graph of $y = |ax + b|$ is identical to the graph of $y = ax + b$ when $ax + b \geqslant 0$ but is given by its reflection in the x-axis when $ax + b < 0$.

For example, compare the graphs of $y = x - 5$ and $y = |x - 5|$.

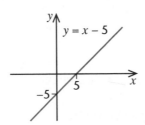 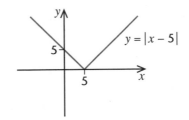

Transforming graphs

The graph of some new function can often be obtained from the graph of a known function by applying a transformation. A summary of the standard transformations is given in the table.

Known function	New function	Transformation
$y = f(x)$	$y = f(x) + a$	Translation through a units parallel to y-axis.
	$y = f(x - a)$	Translation through a units parallel to x-axis.
	$y = af(x)$	One-way stretch with scale factor a parallel to the y-axis.
	$y = f(ax)$	One-way stretch with scale factor $\frac{1}{a}$ parallel to the x-axis.

You may need to apply a combination of transformations in some cases.

Example   The diagram shows the graph of the function $y = f(x)$ where

$$f(x) = 1 - |x - 2| \quad \text{for} \quad 1 \leqslant x \leqslant 3.$$

Use the same axes to show:
(a) $y = f(x) + 1$
(b) $y = f(x + 1)$
(c) $y = 2f(x)$
(d) $y = f(2x)$

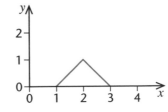

(a) 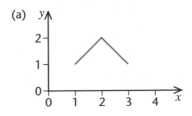   (b)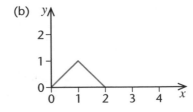

(c) 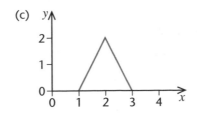   (d)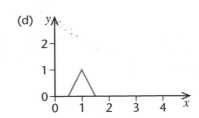

Simultaneous equations

AQA   P1
EDEXCEL P1
OCR   P1
WJEC   P1
NICCEA   P1

The substitution method of solving pairs of linear simultaneous equations can also be applied when one of the equations is a quadratic.

Example Find the co-ordinates of the points where the line $y = 2x + 3$ intersects the circle $x^2 + y^2 = 26$.

The co-ordinates at the points of intersection satisfy both equations simultaneously.

Substitute for y in the quadratic equation: $x^2 + (2x + 3)^2 = 26.$

Now remove the brackets: $x^2 + 4x^2 + 12x + 9 = 26.$

Simplify and write in the form $ax^2 + bx + c = 0$: $5x^2 + 12x - 17 = 0.$

Use the formula to find x: $x = \dfrac{-12 \pm \sqrt{144 + 340}}{10}.$

$x = 1$ or $x = -3.4.$

Substitute the x values in the linear equation to find the corresponding y values: $y = 5$ or $y = -3.8.$

As a check:
$1^2 + 5^2 = 26$
and
$(-3.4)^2 + (-3.8)^2 = 26.$

The co-ordinates at the points of intersection are $(1,5)$ and $(-3.4, -3.8)$.

Identities and equations

AQA   P1
EDEXCEL P1
OCR   P1
WJEC   P1
NICCEA   P1

Identities and equations are closely related but there is an important difference. To make the point consider these two statements.

(1) $x^2 = 3x - 2$

(2) $(x + 5)(x - 5) = x^2 - 25.$

The first statement is an **equation** because it is only true for particular values of x.
The second statement is an **identity** because it is true for *all* values of x.
Depending on the context, the symbol $\equiv$ may be used for an identity in place of $=$ to emphasise its significance.

Example Find the value of A, B and C such that:

$$2x^2 - 9x - 15 \equiv Ax(x + 3) + B(x + 3)^2 + C(x^2 + 1).$$

Since this works for all values of x, you can substitute values of your choice to make simultaneous equations involving A, B and C.

If you choose the values carefully then the equations will be easy to solve:

This makes $x + 3 = 0$ so the terms involving A and B disappear.

Substituting $x = -3$ gives:

$$30 = 10C \implies C = 3.$$

Substituting $x = 0$ gives:

$$-15 = 9B + C \quad \text{but } C = 3 \quad \text{so } -15 = 9B + 3 \Rightarrow B = -2.$$

Sometimes, you run out of convenient values to use. Another approach that works for identities is to match the coefficients of particular terms on each side.

Equating the coefficients of x^2 gives:

$$2 = A + B + C, \quad \text{so } 2 = A - 2 + 3 \Rightarrow A = 1.$$

Polynomials and the factor theorem

AQA	P2
EDEXCEL	P1
OCR	P2
WJEC	P2
NICCEA	P1

This is much simpler than it looks.

A quadratic is one example of a **polynomial**. In general, a polynomial takes the form:

$$a_n x^n + a_{n-1} x^{n-1} + a_{n-2} x^{n-2} + \dots + a_0,$$

where $a_n, a_{n-1}, \dots a_0$ are constants and n is a positive whole number.

For example, $x^4 - 2x^3$ is a polynomial. In this case $a_4 = 1$, $a_3 = -2$ and a_2, a_1, a_0 are all zero. The **degree** of a polynomial is the highest power of x that it includes, so the degree of $x^4 - 2x^3$ is 4. A quadratic is a polynomial of degree 2, a cubic is a polynomial of degree 3 and so on.

Function notation is really useful here. Think about how much more difficult it would be to define the factor theorem without this notation.

One important result that holds for a polynomial of any order is the **factor theorem**. This states that if $f(x)$ is a polynomial and $f(a) = 0$ then $(x - a)$ is a factor of $f(x)$.

Example A polynomial is given by $f(x) = 2x^3 + 13x^2 + 13x - 10$.

(a Find the value of $f(2)$ and $f(-2)$.

(b) State one of the factors of $f(x)$.

(c) Factorise $f(x)$ completely.

(a) $f(2) = 2(2^3) + 13(2^2) + 13(2) - 10 = 84$

$f(-2) = 2(-2)^3 + 13(-2)^2 + 13(-2) - 10 = 0.$

(b) Since $f(-2) = 0$, $(x + 2)$ is a factor of $f(x)$ by the factor theorem.

$f(x)$ is a cubic so the linear factor $(x + 2)$ must be combined with a quadratic.

(c) $2x^3 + 13x^2 + 13x - 10 \equiv (x + 2)(ax^2 + bx + c).$

This is an **identity** and so the values of a, b and c may be found by comparing coefficients.

Comparing the x^3 terms gives: $\qquad a = 2.$

Comparing the x^2 terms gives: $\qquad 13 = 2a + b \Rightarrow b = 9.$

Comparing the constant terms gives: $\qquad c = -5.$

It follows that: $\qquad f(x) = (x + 2)(2x^2 + 9x - 5).$

Factorising the quadratic part in the usual way gives $f(x) = (x + 2)(2x - 1)(x + 5).$

Inequalities

AQA   P1
EDEXCEL P1
OCR   P1
WJEC   P2
NICCEA   P1

Linear inequalities can be solved by rearrangement in much the same way as linear equations. However, care must be taken to reverse the direction of the inequality when multiplying or dividing by a negative.

Example Solve the inequality $8 - 3x > 23$.

Subtract 8 from both sides: $\qquad -3x > 15$
Divide both sides by -3: $\qquad x < -5$.

An inequality which has x on both sides is treated like the corresponding equation.

Example Solve the inequality $5x - 3 > 3x - 10$.

Subtract $3x$ from both sides: $\quad 2x - 3 > -10$
Add 3 to both sides: $\qquad\qquad 2x > -7$
Divide both sides by 2: $\qquad\quad x > -3.5$.

Quadratic inequalities are solved in a similar way to quadratic equations but a sketch graph is often helpful at the final stage.

Example Solve the inequality $x^2 - 3x + 2 < 0$.

Factorise the quadratic expression: $(x - 1)(x - 2) < 0$.

> $y = 0$ when $x = 1$ or when $x = 2$. The graph must cross the x-axis at these points.

Sketch the graph of $y = (x - 1)(x - 2)$:

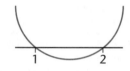

The graph shows that $(x - 1)(x - 2) < 0$ between 1 and 2.

It follows that $x^2 - 3x + 2 < 0$ when $1 < x < 2$.

If the quadratic expression cannot be factorised then the formula may be used to find the points of intersection.

Example Solve the inequality $x^2 + 2x - 5 > 0$.

> $\sqrt{24} = \sqrt{4 \times 6} = 2\sqrt{6}$.

Use the formula to solve $x^2 + 2x - 5 = 0$. $\quad x = \dfrac{-2 \pm \sqrt{24}}{2} = \dfrac{-2 \pm 2\sqrt{6}}{2}$

Simplify the result. $\qquad x = -1 \pm \sqrt{6}$

Sketch the graph of $y = x^2 + 2x - 5$:

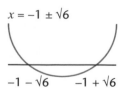

From the sketch, $x^2 + 2x - 5 > 0$ when $x < -1 - \sqrt{6}$ or when $x > -1 + \sqrt{6}$.

Progress check

1 Simplify each of these without using a calculator:

    (a) $25^{\frac{3}{2}}$ (b) $9^{-\frac{1}{2}}$.

2 Simplify these surd expressions:

    (a) $\sqrt{72}$ (b) $(1+\sqrt{3})(1-\sqrt{3})$.

3 A polynomial is given by $f(x) = 3x^3 + 11x^2 - 6x - 8$.

    (a) Find the value of $f(-1)$ and $f(1)$.
    (b) Factorise $f(x)$ completely.
    (c) Solve $f(x) = 0$.
    (d) Sketch the graph of $y = f(x)$.

4 (a) Solve $2x^2 + x - 21 = 0$ by factorising.
    (b) Solve $x^2 - 6x + 7 = 0$ by completing the square.
       Give the roots in surd form.
    (c) Solve $3x^2 + 11x - 2 = 0$ using the quadratic formula.
       Give the roots to 2 decimal places.

5 Express $x^2 - 8x + 17$ in the form $(x-a)^2 + b$ and hence write down its minimum value.

6 Sketch the graph of $y = (x+3)(x-2)^2$.

7 The diagram shows the graph of $y = f(x)$ for $0 \leqslant x \leqslant 2$.
Sketch the graph of:

    (a) $y = f(x-2)$
    (b) $y = f(2x)$.

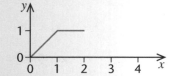

8 Solve these equations simultaneously.
$$x + 2y = 6$$
$$4y^2 - 5x^2 = 36.$$

9 Find the values of A, B and C.
$$3x^3 - 16x - 6 \equiv Ax(x-1)(x+2) + B(x-1)^2 + C(4x^2+3).$$

10 Solve these inequalities.

    (a) $3x - 7 > 5x + 11$ (b) $2x^2 - 3x - 5 \leqslant 0$.

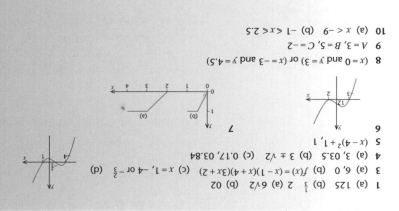

10 (a) $x > -9$ (b) $-1 < x \leqslant 2.5$
9 $A = 3, B = -5, C = -2$
8 $(x = 0$ and $y = 3)$ or $(x = -3$ and $y = 4.5)$
5 $(x-4)^2 + 1$, 1
4 (a) $3, 03.5$ (b) $3 \pm \sqrt{2}$ (c) $0.17, 03.84$
3 (a) $6, 0$ (b) $f(x) = (x-1)(x+4)(3x+2)$ (c) $x = 1, -4$ or $-\frac{2}{3}$ (d)
1 (a) 125 (b) $\frac{1}{3}$ **2** (a) $6\sqrt{2}$ (b) 02

1.2 Coordinate geometry

After studying this section you should be able to:

- *recognise equations of straight lines in various forms*
- *construct the equation of a straight line parallel to a given line and passing through a given point*
- *construct the equation of a straight line perpendicular to a given line and passing through a given point*
- *construct the equation of a straight line passing through two given points*
- *find the coordinates of the mid-point of two given points*

LEARNING SUMMARY

Straight lines

AQA	P1
EDEXCEL	P1
OCR	P1
WJEC	P1
NICCEA	P1

The gradient of a vertical line is undefined.

The general equation of a straight line is $ax + by + c = 0$. The equation of any straight line can be written in this form.

For example, the line $x + y = 5$ corresponds to $a = 1$, $b = 1$ and $c = -5$.

Straight lines, apart from those parallel to the y-axis, can be written in the form $y = mx + c$. This is known as **gradient–intercept** form because the gradient (m) and the y-intercept (c) are clearly identified in the equation. This makes it easy to construct an equation when the gradient and intercept are known.

For example, the line with gradient 4 crossing the y-axis at -5 has equation $y = 4x - 5$.

> Straight lines that are **parallel** must have the **same gradient**.
>
> **KEY POINT**

Example Find the equation of the straight line parallel to $y = 3x - 5$ and passing through the point (4, 2).

The line must have gradient 3 and so it can be written in the form $y = 3x + c$.

This equation must be satisfied at the point where $x = 4$ and $y = 2$.

Substituting $x = 4$ and $y = 2$ gives $2 = 3 \times 4 + c \implies c = -10$.

The required equation is $y = 3x - 10$.

This result is used frequently to find the equation of a tangents or a normal to a curve at a given point.

The equation of a straight line with gradient m and passing through the point (x_1, y_1) can be written as $y - y_1 = m(x - x_1)$.

Using this in the example above gives $y - 2 = 3(x - 4)$. The equation is acceptable in this form but it can be rearranged to give $y = 3x - 10$ as before.

Example Find the equation of the straight line passing through the points (−1, 5) and (3, −2).

One approach is to use the form $y = mx + c$ to produce a pair of simultaneous equations:

Substituting $x = -1$ and $y = 5$ gives $\qquad$ $5 = -m + c$ $\qquad$ (1)
Substituting $x = 3$ and $y = -2$ gives $\qquad$ $-2 = 3m + c$ $\qquad$ (2)

(2) – (1) gives: $\qquad\qquad\qquad\qquad$ $-7 = 4m \Rightarrow m = \dfrac{-7}{4}$.

Substituting for m in (1) gives: $\qquad\qquad$ $5 = \dfrac{7}{4} + c \Rightarrow c = \dfrac{13}{4}$.

The equation of the line is $y = \dfrac{-7}{4}x + \dfrac{13}{4}$. This is the same as $4y + 7x = 13$.

An alternative approach is to find the gradient directly and then use the form $y - y_1 = m(x - x_1)$.

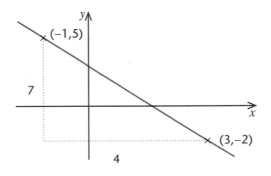

From the diagram, the line has gradient $\dfrac{-7}{4}$.

> You can use either of the points (–1, 5) or (3, –2).

The equation of the line is $y - 5 = \dfrac{-7}{4}(x + 1)$.

To show that this is the same as the previous result (you don't *need* to do this):

Multiply both sides of the equation by 4: $\qquad$ $4y - 20 = -7(x + 1)$.
Expand the brackets: $\qquad\qquad\qquad\qquad$ $4y - 20 = -7x - 7$.
Rearrange to give the previous result: $\qquad$ $4y + 7x = 13$.

An equation of the form $ax + by + c = 0$ can be rearranged into gradient–intercept form provided that $b \neq 0$. This becomes $y = -\frac{a}{b}x - \frac{c}{b}$ and shows that parallel lines may be produced by keeping a and b fixed and allowing c to change.

> Note that $4x - 6y = 7$ is equivalent to $2x - 3y = 3.5$.

For example, the lines $2x - 3y = 5$, $2x - 3y = -2$, $4x - 6y = 7$ are all parallel.

> **KEY POINT**
>
> When two straight lines are **perpendicular**, the **product** of their **gradients** is **–1**.

Example Find the equation of the straight line perpendicular to the line $4x + 3y = 12$ and passing through the point (2,5).

Rearrange the equation into gradient–intercept form: $\qquad$ $y = -\frac{4}{3}x + 4$.
If the gradient of the required line is m then: $\qquad\qquad$ $m \times -\frac{4}{3} = -1 \Rightarrow m = \frac{3}{4}$.
Using the form $y - y_1 = m(x - x_1)$ gives: $\qquad\qquad$ $y - 5 = \frac{3}{4}(x - 2)$.

31

If P has coordinates (x_1, y_1) and Q has coordinates (x_2, y_2) then the mid-point of PQ has coordinates $\left(\dfrac{x_1 + x_2}{2}, \dfrac{y_1 + y_2}{2}\right)$.

For example, the mid-point of $(-3,1)$ and $(5,7)$ is $\left(\dfrac{-3+5}{2}, \dfrac{1+7}{2}\right) = (1,4)$.

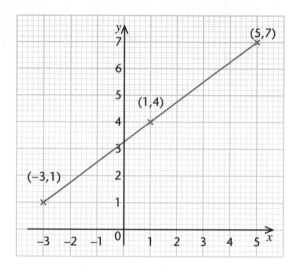

Progress check

The parallel line can be written in the form $3x + 2y = C$. To find C just substitute the coordinates $(5,4)$ for x and y.

1 The equation of a straight line is $y = 2x - 7$.

  (a) Write down the equation of a parallel line crossing the y-axis at 5.

  (b) Find the equation of a parallel line passing through the point $(4,9)$.

2 Find the equation of the straight line joining the points $(-2,-9)$ and $(1,3)$.

3 The equation of a straight line is $3x + 2y = 11$.

  (a) Find the equation of a parallel line passing through the point $(5,4)$.

  (b) Find the equation of a perpendicular line passing through the point $(-2,6)$.

4 Find the co-ordinates of the mid-point of $(-9,3)$ and $(-5,1)$.

5 P is the point $(-1,-4)$ and Q is the point $(5,-1)$. Find the equation of the line perpendicular to PQ and passing through its mid-point.

1 (a) $y = 2x + 5$ (b) $y = 2x + 1$
2 $y = 4x - 1$
3 (a) $3x + 2y = 23$ (b) $y - 6 = \frac{2}{3}(x + 2)$ or $3y - 2x - 22 = 0$.
4 $(-7, 2)$
5 $y = -2x + 1.5$

1.3 Sequences and series

After studying this section you should be able to:

- understand the different types of sequence, including the notation and formulae used to describe them
- recognise arithmetic and geometric progressions and calculate the sum of their series
- determine whether a geometric progression with an infinite number of terms has a finite sum and calculate its value when it exists

LEARNING SUMMARY

Sequences

AQA	P1
EDEXCEL	P1
OCR	P2
WJEC	P1
NICCEA	P2

A list of numbers in a particular order, that follow some rule for finding later values, is called a **sequence**. Each number in a sequence is called a **term**, and terms are often denoted by $u_1, u_2, u_3, ..., u_n, ...$.

One way to define a sequence is to give a formula for the nth term such as $u_n = n^2$. Substituting the values $n = 1, 2, 3, 4, ...$ produces the sequence $1, 4, 9, 16, ...$ and the value of any particular term can be calculated by substituting its position number into the formula. For example, in this case, the 50th term is $u_{50} = 50^2 = 2500$.

Another way to define a sequence is to give a starting value together with a rule that shows the connection between successive terms. This is sometimes called a **recursive definition**. For example $u_1 = 5$ and $u_{n+1} = 2u_n$ defines the sequence $5, 10, 20, 40, ...$. The rule $u_{n+1} = 2u_n$ is an example of a **recurrence relation**.

Two special sequences are the **arithmetic progression** (A.P.) and the **geometric progression** (G.P.). In an A.P. successive terms have a common difference, e.g. $1, 4, 7, 10, ...$. The first term is denoted by a and the common difference is d. With this notation, the definition of an A.P. may now be given as $u_1 = a$, $u_{n+1} = u_n + d$.

The terms of an A.P. take the form $a, a + d, a + 2d, a + 3d, ...$ and the nth term is given by $u_n = a + (n - 1)d$.

In a G.P. successive terms are connected by a **common ratio**, e.g. $3, 6, 12, 24, ...$. The definition of a G.P. may be given as $u_1 = a$, $u_{n+1} = ru_n$, where r is the common ratio. The terms of a G.P. take the form $a, ar, ar^2, ar^3, ...$ and the nth term is given by $u_n = ar^{n-1}$.

Series

AQA	P1
EDEXCEL	P1
OCR	P2
WJEC	P1
NICCEA	P2

A series is formed by adding together the terms of a sequence. The use of sigma notation can greatly simplify the way that series are written. For example, the series $1^2 + 2^2 + 3^2 + ... + n^2$ may be written as $\sum_{i=1}^{n} i^2$. The sum of the first n terms of a series is often denoted by S_n and so $S_n = u_1 + u_2 + u_3 + ... + u_n = \sum_{i=1}^{n} u_i$.

The **sum of an A.P.** is given by $S_n = a + (a + d) + (a + 2d) + ... + (a + (n - 1)d)$.

This may also be written as $S_n = l + (l - d) + (l - 2d) + ... + (l - (n - 1)d)$, where l is the last term.

Adding the two versions gives $2S_n = (a + l) + (a + l) + (a + l) + ... (a + l)$
$$= n(a + l).$$

> You need to know how to establish the general result.

This gives the general result: $S_n = \dfrac{n}{2}(a+l)$.

Substituting $l = a + (n-1)d$ gives the alternative form of the result:

$$S_n = \frac{n}{2}(2a + (n-1)d).$$

A special case is the sum of the first n natural numbers:

$$1 + 2 + 3 + \ldots + n = \frac{n}{2}(n+1).$$

For example, the sum of the first 100 natural numbers is $\frac{100}{2}(100+1) = 5050$.

When finding the sum of an A.P. you need to select the most appropriate version of the formula to suit the information.

Example Find the sum of the first 50 terms of the series $15 + 18 + 21 + 24 + \ldots$.

In this series, $a = 15$, $d = 3$ and $n = 50$.

Using $S_n = \dfrac{n}{2}(2a + (n-1)d)$ gives $S_{50} = \dfrac{50}{2}(2 \times 15 + 49 \times 3) = 4425$.

The sum of a G.P. is given by $\qquad S_n = a + ar + ar^2 + ar^3 + \ldots + ar^{n-1}$.

Multiplying throughout by r gives $rS_n = ar + ar^2 + ar^3 + \ldots + ar^{n-1} + ar^n$.

Subtracting gives: $\qquad\qquad S_n - rS_n = a - ar^n$

$$\Rightarrow S_n(1-r) = a(1-r^n)$$

> You need to know how to establish this result.

So the sum of the first n terms of a G.P. is $S_n = \dfrac{a(1-r^n)}{1-r}$.

If $r > 1$ then it is more convenient to use the result in the form $S_n = \dfrac{a(r^n - 1)}{r-1}$.

Example Find the sum of the first 20 terms of the series $8 + 12 + 18 + 27 + \ldots$ to the nearest whole number.

In this series, $a = 8$, $r = 1.5$ and $n = 20$.

This gives $S_{20} = \dfrac{8(1.5^{20} - 1)}{1.5 - 1} = 53188.107\ldots = 53188$ to the nearest whole number.

Provided that $|r| < 1$ the sum of a G.P. converges to $\dfrac{a}{1-r}$ as n tends to infinity.

For example, $1 + \dfrac{1}{2} + \dfrac{1}{4} + \dfrac{1}{8} + \ldots = \dfrac{1}{1 - \frac{1}{2}} = 2$.

Progress check

1. Write down the first five terms of the sequence given by:
   (a) $u_n = 2n^2$ (b) $u_1 = 10$, $u_{n+1} = 3u_n + 2$.

2. Find the 20th term of an A.P. with first term 7 and common difference 5.

3. Find the sum of the first 1000 natural numbers.

4. Find the sum of the first 20 terms of $12 + 15 + 18.75 + \ldots$ to 4 s.f.

5. Explain why the series $10 + 9 + 8.1 + 7.29 + \ldots$ is convergent and find the value of its sum to infinity to 5 s.f.

3 500 500   4 4115   5 $|r| = 0.9 < 1$, 100.

2 102

1 (a) 2, 8, 18, 32, 50   (b) 10, 32, 98, 296, 890

1.4 Trigonometry

After studying this section you should be able to:

- understand the properties of the sine, cosine and tangent functions
- sketch the graphs of trigonometric functions
- use identities to simplify expressions and solve equations

The sine cosine and tangent of 30°, 45°, and 60° may be expressed exactly as shown below. The results are based on Pythagoras' theorem.

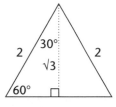 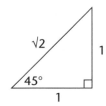

$$\sin 30° = \frac{1}{2} \qquad \sin 60° = \frac{\sqrt{3}}{2} \qquad \sin 45° = \frac{1}{\sqrt{2}}$$

$$\cos 30° = \frac{\sqrt{3}}{2} \qquad \cos 60° = \frac{1}{2} \qquad \cos 45° = \frac{1}{\sqrt{2}}$$

$$\tan 30° = \frac{1}{\sqrt{3}} \qquad \tan 60° = \sqrt{3} \qquad \tan 45° = 1.$$

You need to be able to sketch the graphs of the sine, cosine and tangent functions and to know their special properties.

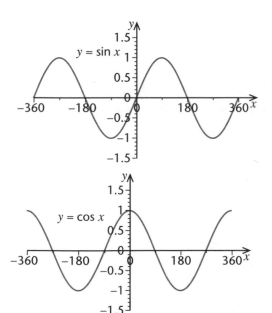

$\sin x$ is defined for any angle and always has a value between -1 and 1. It is a **periodic function** with period 360°.

The graph has **rotational symmetry** of order 2 about every point where it crosses the x-axis.

It has **line symmetry** about every vertical line passing through a vertex.

$\cos x \equiv \sin(x + 90°)$ so the graph of $y = \cos x$ can be obtained by translating the sine graph 90° to the left.

It follows that $\cos x$ is also a periodic function with period 360° and has the corresponding symmetry properties.

$$\tan x \equiv \frac{\sin x}{\cos x}.$$

$\tan x$ is undefined whenever $\cos x = 0$ and approaches $\pm\infty$ near these values. It is a periodic function with period 180°.

The graph has rotational symmetry of order 2 about 0°, $\pm 90°$, $\pm 180°$, $\pm 270°$, ...

The graphs of more complex trigonometric functions can often be produced by applying transformations to one of the basic graphs.

Example   Sketch the graph of $y = \sin(2x + 30) + 1$ for $0° \leqslant x \leqslant 360°$.

It is helpful to think about building the transformations in stages.

Basic function	Transformations		
	Apply a one-way stretch with scale factor $\frac{1}{2}$ parallel to the x-axis.	Now translate the curve 30° to the left.	Finally translate the curve 1 unit up.
$y = \sin x$	$y = \sin(2x)$	$y = \sin(2x + 30)$	$y = \sin(2x + 30) + 1$

The graph of $y = \sin(2x + 30°) + 1$ may now be produced by applying these transformations in order starting from the graph of $y = \sin x$.

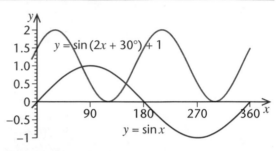

Identities and equations

AQA	P1
EDEXCEL	P1
OCR	P1
WJEC	P1
NICCEA	P1

> **KEY POINT**
>
> The identities $\tan x \equiv \dfrac{\sin x}{\cos x}$ and $\sin^2 x + \cos^2 x \equiv 1$ are very important in terms of simplifying expressions and solving equations.

Example   (a) Simplify $1 - (\sin x - \cos x)^2$

(b) Solve $2 \cos x = 5 \sin x$ for $-360° \leqslant x \leqslant 360°$.

With practice you learn to create opportunities to make use of the identities.

(a) $1 - (\sin x - \cos x)^2 \equiv 1 - \sin^2 x + 2 \sin x \cos x - \cos^2 x$
$$\equiv 1 - (\sin^2 x + \cos^2 x) + 2 \sin x \cos x$$
$$\equiv 2 \sin x \cos x.$$

(b) $2 \cos x = 5 \sin x \Rightarrow 2 = \dfrac{5 \sin x}{\cos x}$
$$\Rightarrow 2 = 5 \tan x$$
$$\Rightarrow \tan x = \tfrac{2}{5}.$$

One solution is $x = \tan^{-1}(\tfrac{2}{5}) = 21.8°$ to 1 d.p.

To find the other solutions in the interval $-360° \leqslant x \leqslant 360°$ you can look at the sketch of $y = \tan x$.

The tan function has period 180° so the graph repeats itself every 180°.

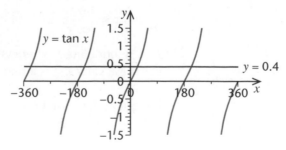

The graph shows that there are four solutions in the required interval. These values are 180° apart.

The solutions are:
201.8°, 21.8°, −158.2°, −338.2°.

You may be required to solve trigonometric equations in a variety of forms.

Example   Solve $\sin(2x - 30°) = 0.6$ for $0° \leqslant x \leqslant 360°$.

> First find a value of $2x - 30°$.

One value of $2x - 30°$ is $\sin^{-1}(0.6) = 36.86...°$.
You need to adjust the interval to find where the values of $2x - 30°$ must lie.

$$0° \leqslant x \leqslant 360° \Rightarrow 0° \leqslant 2x \leqslant 720° \Rightarrow -30° \leqslant 2x - 30° \leqslant 690°.$$

Now look at the graph of the sine function in this interval.

> This is the graph of $y = \sin X$ for $-30° \leqslant X \leqslant 690°$ where $X = 2x - 30°$.
> In other words, draw the sine curve as normal for the interval and then interpret the variable as $2x - 30°$.

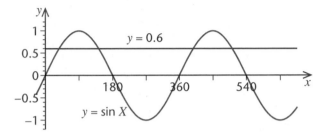

The graph shows that there are four values of $2x - 30°$ in the interval that make the equation work.

Using the symmetry of the curve, the values of $2x - 30°$ are:
$36.86...°$, $180° - 36.86...°$, $360° + 36.86...°$, $540° - 36.86...°$.

This gives:
$2x - 30° = 36.86...° \Rightarrow x = 33.4°$; $2x - 30° = 143.14...° \Rightarrow x = 86.6°$.
$2x - 30° = 396.86...° \Rightarrow x = 213.4°$; $2x - 30° = 503.14...° \Rightarrow x = 266.6°$.

Radians

AQA	P2
EDEXCEL	P1
OCR	P2
WJEC	P1
NICCEA	P1

Angles can also be measured in **radians** and this makes it much easier to deal with trigonometric functions when using **calculus**.

1 radian $\approx 57.3°$ this may be written as $1^c \approx 57.3°$. However, the symbol for radians is not normally written when the angle involves π. The following results are useful to remember: $\pi = 180°$, $\dfrac{\pi}{2} = 90°$, $\dfrac{\pi}{4} = 45°$, $\dfrac{\pi}{3} = 60°$, $\dfrac{\pi}{6} = 30°$.

All of the results that you know in degrees also apply in radians, e.g. $\sin \dfrac{\pi}{6} = 0.5$.

Progress check

1 Describe the transformations needed to turn the graph of $y = \cos x$ into the graph of:

(a) $\cos 2x$ (b) $3 \cos x$ (c) $3 \cos(2x) - 1$.

2 Show that $(\sin x + \cos x)^2 + (\sin x - \cos x)^2 \equiv 2$.

3 Solve $8 \sin x = 3 \cos x$ for $-360° \leqslant x \leqslant 360°$.

4 Solve $\cos(3x + 20°) = 0.6$ for $0° \leqslant x \leqslant 180°$.

4 $11.0°$, $95.6°$, $131.1°$

3 $200.6°$, $20.6°$, $-159.4°$, $-339.4°$

2 $\sin^2 x + 2 \sin x \cos x + \cos^2 x + \sin^2 x - 2 \sin x \cos x + \cos^2 x \equiv 2(\sin^2 x + \cos^2 x) \equiv 2$

1 (a) One-way stretch with scale factor 0.5 parallel to the x-axis.
(b) One-way stretch with scale factor 3 parallel to the y-axis.
(c) One-way stretch with scale factor 0.5 parallel to the x-axis, followed by a one-way stretch with scale factor 3 parallel to the y-axis, followed by a translation of 1 unit downwards.

1.5 Differentiation

After studying this section you should be able to:

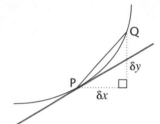

- *use differentiation to find the exact value of the gradient of a curve*
- *find the equation of the tangent and normal to a curve at a given point*
- *determine whether a function is increasing or decreasing in an interval*
- *locate turning points and use the second derivative to distinguish between them*

The **gradient** of a curve changes continuously along its length. Its value at any point P is given by the gradient of the tangent to the curve at P. The gradient may be found approximately by drawing.

The exact value of the gradient is found by **differentiation**. This is the **limit** of the gradient of PQ as Q moves towards P.

Gradient of curve at P $\approx \dfrac{\delta y}{\delta x}$.

As Q moves towards P, $\delta x \rightsquigarrow 0$ and $\dfrac{\delta y}{\delta x} \rightsquigarrow \dfrac{dy}{dx}$.

> **KEY POINT**
>
> $\dfrac{dy}{dx}$ is the **gradient function** and represents the **derivative** of y with respect to x.
>
> The graph of $y = kx^n$ has gradient function $\dfrac{dy}{dx} = nkx^{n-1}$.

Example Find the gradient of the curve $y = 3x^2$ at the point P (5,75).

$\dfrac{dy}{dx} = 2 \times 3x^1 = 6x.$ At P: $x = 5$ so the gradient is $6 \times 5 = 30$.

Function notation may also be used for derivatives. If $y = f(x)$ then $\dfrac{dy}{dx} = f'(x)$. The notation is useful for stating some of the rules of differentiation.

> **KEY POINT**
>
> If $y = f(x) \pm g(x)$ then $\dfrac{dy}{dx} = f'(x) \pm g'(x)$

Example $f(x) = x^3 + 5x + 2$. Find (a) $f'(x)$ (b) $f'(4)$.

Differentiate each term separately.
Remember that $x^0 = 1$.
When you differentiate a constant, the result is zero.

(a) $f'(x) = 3x^2 + 5$ (b) $f'(4) = 3 \times 4^2 + 5 = 53$.

You often need to express a function in the right form before you can differentiate it.

Example Differentiate: (a) $f(x) = \sqrt{x}$ (b) $f(x) = (x + 5)(x - 3)$ (c) $f(x) = \dfrac{x^3 + 1}{x^2}$.

(a) $\sqrt{x}$ has to be written as a power of x to use the rule $f'(x) = nkx^{n-1}$.

$\sqrt{x} = x^{\frac{1}{2}}$ so $f(x) = x^{\frac{1}{2}}$ and $f'(x) = \frac{1}{2}x^{-\frac{1}{2}}$.

> You cannot differentiate a product term by term.

(b) The brackets must first be removed and *then* you can differentiate term by term.

$f(x) = (x - 5)(x + 3) = x^2 - 2x - 6$.

$f'(x) = 2x - 2$.

> You cannot differentiate a quotient term by term.

(c) Divide $x^3 + 1$ by x^2 first giving $f(x) = x + x^{-2}$.

Now, $f'(x) = 1 - 2x^{-3} = 1 - \dfrac{2}{x^3}$.

Tangents and normals

AQA	P3
EDEXCEL	P1
OCR	P1
WJEC	P3
NICCEA	P2

You can find the gradient of the tangent to a curve at a point by differentiation. Then you can use the techniques described in the Coordinate Geometry section to find the **equation of the tangent** and the **equation of the normal** at the given point.

Example Find the equation of the tangent and the normal to the curve $y = x^3 - 4x$ at the point (2,0).

Differentiate the equation of the curve to give $\dfrac{dy}{dx} = 3x^2 - 4$.

The gradient of the tangent at (2,0) is $3 \times 2^2 - 4 = 8$.
Using $y - y_1 = m(x - x_1)$ gives the equation of the tangent as $y = 8(x - 2)$.

The gradient of the normal is $-\frac{1}{8}$. Using $y - y_1 = m(x - x_1)$ again gives the equation of the normal as $y = -\frac{1}{8}(x - 2)$. You could rearrange this to give $x + 8y = 2$.

Curve sketching

AQA	P2
EDEXCEL	P1
OCR	P1
WJEC	P1
NICCEA	P1

The gradient function gives information about the behaviour of a curve.

$f'(x) > 0$ at A and the function is **increasing**

$f'(x) < 0$ at C and the function is **decreasing**

$f'(x) = 0$ at B and D and the function is neither increasing nor decreasing. These points are called **stationary points**. B is at a local **maximum** and D is at a **local minimum**.

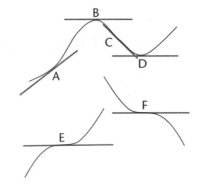

E and F show a different type of stationary point called a called a **stationary point of inflexion**.

One way to distinguish between the three types of stationary point is to look at the sign of the derivative on either side of the point.

> The first two types of stationary point are also known as **turning points**.

- At a local maximum the sign of the derivative changes from positive to negative.
- At a local minimum the sign of the derivative changes from negative to positive.
- At a stationary point of inflexion there is no change of sign.

The second derivative

AQA	P2
EDEXCEL	P1
OCR	P1
WJEC	P1
NICCEA	P1

Starting from $\dfrac{dy}{dx} = f'(x)$ and differentiating again gives the **second derivative**. This is written as $\dfrac{d^2y}{dx^2} = f''(x)$. The second derivative can give information about the nature of any stationary points.

At a stationary point:

- $f''(x) > 0 \Rightarrow$ the point is a local minimum
- $f''(x) < 0 \Rightarrow$ the point is a local maximum.

But, if $f''(x) = 0$ then this gives no further information.

Example Find the stationary points of the curve $y = \dfrac{x^3}{3} - \dfrac{3}{2}x^2 + 2x - 1$ and use the second derivative to distinguish between them.

First differentiate the equation, then solve $\dfrac{dy}{dx} = 0$ to locate the stationary points.

$$\frac{dy}{dx} = x^2 - 3x + 2.$$

At a stationary point $\dfrac{dy}{dx} = 0 \Rightarrow x^2 - 3x + 2 = 0$

$$\Rightarrow (x - 1)(x - 2) = 0$$
$$\Rightarrow x = 1 \text{ or } x = 2.$$

When $x = 1$, $y = \frac{1}{3} - \frac{3}{2} + 2 - 1 = -\frac{1}{6}$.

When $x = 2$, $y = \frac{8}{3} - 6 + 4 - 1 = -\frac{1}{3}$.

The stationary points are $(1, -\frac{1}{6})$ and $(2, -\frac{1}{3})$.

In this case, the function is a cubic and so the shape of the curve is known. We should expect the first stationary point to be a local maximum and the second to be a local minimum.

Using the second derivative: $f''(x) = 2x - 3$

so $f''(1) = -1 < 0$ giving a local maximum,

and $f''(2) = 1 > 0$ giving a local minimum as expected.

Progress check

1 Differentiate with respect to x.
   (a) $y = 4x^3 - 7x^2 + 3x - 9$ (b) $y = 6\sqrt{x}$.

2 Differentiate with respect to x:
   (a) $f(x) = (3x + 1)(x - 4)$ (b) $f(x) = \dfrac{2x + 1}{x^3}$.

3 Find the gradient of the curve $y = (4x - 1)^2$ at the point $(1, 9)$.

Solve the inequality
$f'(x) < 0$.

4 Find the values of x for which the function $f(x) = x^3 - 6x^2 + 9x - 7$ is decreasing.

5 Find the equation of: (a) the tangent   (b) the normal
   to the curve $y = 12\sqrt{x}$ at the point $(9, 36)$.

6 Locate the stationary points on the curve $y = 2x^3 - 6x^2 - 18x + 5$ and use the second derivative to distinguish between them.

6 Local maximum: $(-1, 15)$; local minimum $(3, -49)$.

5 Tangent: $y = 2x + 18$; normal: $y = -\frac{1}{2}(x) = 36 - (x - 9)$.

4 $1 < x < 3$

3 24

1 (a) $12x^2 - 14x + 3$ (b) $\dfrac{3}{\sqrt{x}}$ 2 (a) $6x - 11$ (b) $-\dfrac{4}{x^3} - \dfrac{x}{x^4}$

1.6 Integration

After studying this section you should be able to:

- find an indefinite integral and understand what it represents
- evaluate a definite integral and apply this to finding the area under a curve

LEARNING SUMMARY

The idea of a **reverse process** is an important one in many areas of mathematics. In **calculus**, the reverse process of differentiation is **integration** and this turns out to be an extremely important process, in its own right, with many powerful applications.

When you differentiate x^n with respect to x you may think of the process involving two stages:

$$x^n \rightarrow \boxed{\text{multiply by the power}} \rightarrow \boxed{\text{reduce the power by 1}} \rightarrow nx^{n-1}$$

This *suggests* that the reverse process is given by:

$$\frac{x^{n+1}}{n+1} \leftarrow \boxed{\text{divide by the power}} \leftarrow \boxed{\text{increase the power by 1}} \leftarrow x^n$$

However, this doesn't give the *complete* picture. The reason is that you can differentiate $x^n + (\text{any constant})$ and still obtain nx^{n-1}. You need to take this into account when you reverse the process.

$\int$ is the symbol for integration and $\mathrm{d}x$ is used to show that the integration is with respect to the variable x.

The result is written as $\displaystyle\int x^n \, \mathrm{d}x = \frac{x^{n+1}}{n+1} + c$ where c is called the **constant of integration**.

Some examples are:

Notice that $\dfrac{1}{\left(\frac{3}{2}\right)} = \dfrac{2}{3}$.

$$\int x^2 \, \mathrm{d}x = \frac{x^3}{3} + c \qquad \int \sqrt{x} \, \mathrm{d}x = \int x^{\frac{1}{2}} \, \mathrm{d}x = \frac{2}{3}x^{\frac{3}{2}} + c$$

$$\int \frac{1}{x^2} \, \mathrm{d}x = \int x^{-2} \, \mathrm{d}x = -x^{-1} + c = -\frac{1}{x} + c.$$

Sums and differences of functions are treated in the same way as in differentiation, by dealing with each term separately. The general rules are given below.

Function notation is useful here but you don't need to state the rules every time you use them.

> $$\int f(x) \pm g(x) \, \mathrm{d}x = \int f(x) \, \mathrm{d}x \pm \int g(x) \, \mathrm{d}x$$
>
> $$\int af(x) \, \mathrm{d}x = a \int f(x) \, \mathrm{d}x \quad \text{(where } a \text{ is a constant)}$$

KEY POINT

For example,

$$\int (3x^2 - 5x + 2) \, \mathrm{d}x = x^3 - 5\frac{x^2}{2} + 2x + c.$$

In some situations you have enough information to find the value of the constant.

Example Find the equation of the curve with gradient $3x^2$ passing through $(2, 5)$.

Using the point $(2, 5)$ gives
$5 = 2^3 + c$
so $c = -3$.

$$\frac{\mathrm{d}y}{\mathrm{d}x} = 3x^2 \implies y = \int 3x^2 \, \mathrm{d}x \implies y = x^3 + c. \quad \text{When } x = 2, y = 5 \implies c = -3.$$

The equation of the curve is $y = x^3 - 3$.

The area under a curve

AQA   P1
EDEXCEL   P1
OCR   P1
WJEC   P1
NICCEA   P1

An integral can take different forms to suit a particular purpose. An integral of the form $\int f(x)\,dx = g(x)$ is called an **indefinite integral** and is useful when the end result is required to be a function. All of the ones looked at so far have been indefinite integrals.

An integral of the form $\int_a^b f(x)\,dx = [g(x)]_a^b$ is called **a definite integral** and has a numerical value given by $g(b) - g(a)$. This represents the area under the curve $y = f(x)$ between the lines $x = a$ and $x = b$.

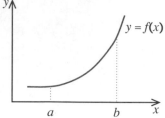

Example Find the area under the curve $y = x^2 + 1$ between $x = 1$ and $x = 3$.

> A constant of integration is **not** used in a definite integral.

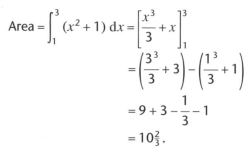

$$\text{Area} = \int_1^3 (x^2 + 1)\,dx = \left[\frac{x^3}{3} + x\right]_1^3$$

$$= \left(\frac{3^3}{3} + 3\right) - \left(\frac{1^3}{3} + 1\right)$$

$$= 9 + 3 - \frac{1}{3} - 1$$

$$= 10\tfrac{2}{3}.$$

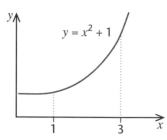

The area between a curve and the y-axis

OCR   P1
NICCEA   P1

The area bounded by the curve $y = f(x)$, the y-axis, and the lines $y = a$ and $y = b$ is given by $\int_a^b x\,dy.$

> In this case, you have to write x in terms of y and integrate with respect to y.

Example Find the area between the curve $y = x^3$ and the y-axis between $y = 1$ and $y = 8$.

Since $y = x^3$, it follows that $x = y^{\frac{1}{3}}$.

The required area is given by

$$\int_1^8 y^{\frac{1}{3}}\,dy = \left[\frac{3}{4} y^{\frac{4}{3}}\right]_1^8 = \frac{3}{4}(8^{\frac{4}{3}} - 1)$$

$$= \frac{3}{4} \times 15 = 11\tfrac{1}{4}.$$

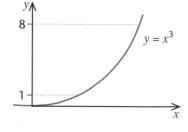

Progress check

1 Integrate each of these with respect to x.

(a) x^4 (b) $6x^2 - 4x + 3$ (c) $\dfrac{1}{\sqrt{x}}$ (d) $\sqrt{x^3}$.

2 Find the value of each of these definite integrals.

(a) $\int_0^5 (x - 2)\,dx$ (b) $\int_4^9 \sqrt{x}\,dx$ (c) $\int_0^6 y^2\,dy$

3 A curve has gradient function $4x^3 + 1$ and passes through the point $(1,9)$. Find the equation of the curve.

4 Find the area under the curve $y = 3\sqrt{x}$ between $x = 1$ and $x = 9$.

4 52

3 $y = x^4 + x + 7$

2 (a) 2.5 (b) $12\tfrac{2}{3}$ (c) 72

1 (a) $\dfrac{x^5}{5} + c$ (b) $2x^3 - 2x^2 + 3x + c$ (c) $2\sqrt{x} + c$

Sample questions and model answers

1

(a) Write down the exact value of 3^{-3}.

'Write down' means that you don't need to show any working to get the marks. But you might still find it useful as a way to clarify your thoughts.

Remember that $a^{-n} = \dfrac{1}{a^n}$

(b) Simplify $\dfrac{(4x^4)^{\frac{3}{2}}}{2\sqrt{x} \times x\sqrt{x}}$.

(a) $3^{-3} = \dfrac{1}{3^3} = \dfrac{1}{27}$.

(b) $\dfrac{(4x^4)^{\frac{3}{2}}}{2\sqrt{x} \times x\sqrt{x}} = \dfrac{4^{\frac{3}{2}}(x^4)^{\frac{3}{2}}}{2x^2}$

$4^{\frac{3}{2}} = (\sqrt{4})^3 = 2^3 = 8$
$(x^4)^{\frac{3}{2}} = x^{4 \times \frac{3}{2}} = x^6$.

$= \dfrac{8x^6}{2x^2}$

$= 4x^4$.

2

(a) Write $2x^2 - 12x + 11$ in the form $a(x+b)^2 + c$.

(b) State the minimum value of $2x^2 - 12x + 11$ and give the value of x where this occurs.

Start by taking out the factor of 2 common to the first two terms.

(c) Solve the equation $2x^2 - 12x + 11 = 0$ and express your answer in surd form.

Keep the 11 separate rather than have a fraction inside the brackets.

(d) Sketch the graph of $y = 2x^2 - 12x + 11$.

Complete the square and simplify.

(a) $2x^2 - 12x + 11 = 2(x^2 - 6x) + 11$

$= 2((x - 3)^2 - 9) + 11$

$= 2(x - 3)^2 - 7$.

The minimum value occurs when the expression in the brackets equals zero.

(b) The minimum value is -7 and this occurs when $x = 3$.

(c) $2x^2 - 12x + 11 = 0$

$\Rightarrow 2(x - 3)^2 - 7 = 0$

$\Rightarrow (x - 3)^2 = \frac{7}{2} = \frac{14}{4}$

$\Rightarrow x - 3 = \pm \frac{1}{2}\sqrt{14}$

$\Rightarrow x = 3 \pm \frac{1}{2}\sqrt{14}$.

(d)

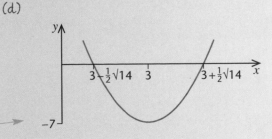

Use the information you have found to position the curve and label the key points.

Sample questions and model answers *(continued)*

3

Solve the equation $5 \sin x - 3 \cos^2 x = 0$ for values of x from 0 to 360°.

Replacing $\cos^2 x$ with $1 - \sin^2 x$ in the equation gives:

> This is a quadratic equation in $\sin x$.

$$5 \sin x - 3(1 - \sin^2 x) = 0$$

$$\Longrightarrow \quad 3 \sin^2 x + 5 \sin x - 3 = 0$$

> You can use the quadratic formula with $a = 3$, $b = 5$ and $c = -3$.

$$\Longrightarrow \quad \sin x = \frac{-5 \pm \sqrt{25 + 36}}{6}$$

$$\Longrightarrow \quad \sin x = 0.46837 \ldots \text{ or } \sin x = -2.135 \ldots \text{ (no solutions)}.$$

> It's a good idea to sketch the graph of the function as a guide to locating all of the solutions.

One solution is given by $\sin^{-1}(0.46837\ldots) = 27.9°$.

The diagram shows that a second solution in the required interval is $180° - 27.9° = 152.1°$.

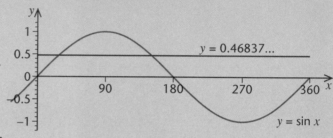

So, $x = 27.0°$ or $x = 152.1°$.

4

(a) Find the equation of the normal to the curve $y = x(x - 2)$ at the point P (2.5, 1.25).

(b) Find the coordinates of the point Q where the normal crosses the x-axis.

(c) Calculate the area shown shaded in the diagram.

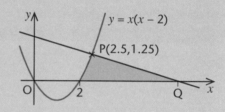

(a) $y = x(x - 2) = x^2 - 2x$

> Multiply out the brackets before differentiating.

$$\Longrightarrow \quad \frac{dy}{dx} = 2x - 2.$$

So the gradient of the tangent at P is $2 \times 2.5 - 2 = 3$ and the gradient of the normal at P is $-\frac{1}{3}$.

> The product of these gradients must be −1 since they are at right angles to each other.

The equation of the normal at P is $y - 1.25 = -\frac{1}{3}(x - 2.5)$.

> This is using the equation of a straight line in the form $y - y_1 = m(x - x_1)$.

(b) At Q, $y = 0 \Longrightarrow -1.25 = -\frac{1}{3}(x - 2.5) = 0$

$$\Longrightarrow \quad x = 6.25$$

So Q has coordinates (6.25, 0).

(c) The shaded area is given by

> The area may be split into two parts: the part under the curve and the remaining triangle.

$$\int_2^{2.5} (x^2 - 2x)\, dx + \frac{1}{2}(6.25 - 2.5) \times 1.25$$

$$= \left[\frac{x^3}{3} - x^2 \right]_2^{2.5} + 2.34375 = 2.635 \text{ to 4 s.f.}$$

Practice examination questions

1 (a) Write down the exact value of $49^{-\frac{1}{2}}$

(b) Find the value of k such that $a^k = \dfrac{a \times \sqrt{a^3}}{\sqrt{a}}$.

2 (a) Given that $8^{x-3} = 4^{x+1}$, find the value of x.

(b) Solve the equation $5x^{-\frac{1}{3}} = x^{\frac{1}{6}}$.

3 (a) Write $x^2 - 4x + 1$ in the form $(x + a)^2 + b$.

(b) Solve $x^2 - 4x + 1 = 0$ and express your answer in surd form.

(c) State the coordinates of the lowest point on the graph of $y = x^2 - 4x + 1$.

(d) Sketch the graph.

4 A function $f(x)$ is defined by:

$$f(x) = x^3 + 4x^2 + kx - 18$$

for all real values of x, where k is a constant.

(a) Given that $(x - 2)$ is a factor of $f(x)$, find the value of k.

(b) Factorise $f(x)$ completely and sketch the graph of $y = f(x)$.

(c) Sketch the graphs of $y = f(x) + 10$ and $y = f(x - 2)$.

5 Find the coordinates of the points A and B at the intersection of these graphs.

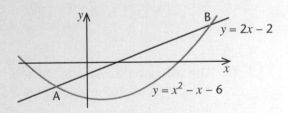

6 (a) Find the values of A, B and C for which

$$4x^2 + 27x + 94 \equiv A(x - 1)^2 + B(x - 1)(x + 4) + C(x + 4)^2.$$

(b) Solve the inequality

$$3x^2 - 8x - 7 < 2x^2 - 3x - 11.$$

Practice examination questions (continued)

7 (a) Find the 8th term of a geometric progression with first term 3 and common ratio 2.

   (b) Find the sum of the first 50 terms of the series $25 + 26.2 + 27.4 + \ldots.$

8 The 10th term of an arithmetic progression is 74 and the sum of the first 20 terms is 1510.
   Find the first term and the common difference.

9 A ball is dropped from a height of 2 m above a hard floor.
   After each bounce, the ball rises to $\frac{3}{4}$ of its previous height.

   (a) How far has it travelled in total when it strikes the floor for the 4th time?

   (b) How far has it travelled in total when it strikes the floor for the nth time?

   (c) Show that the ball never travels more than 14 m in total.

10 Solve the equation $5 \sin x = 8 \cos x$ for values of x in the interval $-360° \leqslant x \leqslant 360$.

11 (a) Write the equation $3 \sec^2 x + \tan x - 6 = 0$ as a quadratic equation in $\tan x$.

   (b) Solve the equation giving the solutions in degrees to the nearest degree for $-180° \leqslant x \leqslant 180°$.

12 (a) Sketch the graph of $y = 2 \sin(x - 30°)$ for $0° \leqslant x \leqslant 360°$.

   (b) Solve the equation $2 \sin(x - 30°) = \sqrt{3}$ for $0° \leqslant x \leqslant 360°$.

13 (a) The diagram shows a straight line passing through the points A$(-1,-5)$ and B$(4,1)$.
   Find its equation in the form $ax + by + c = 0$.

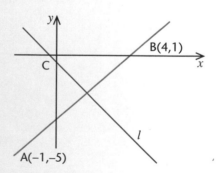

   (b) A second line l, passes through the mid-point of A and B at right angles to AB.

   Find its equation in the form $ax + by + c = 0$.

   (c) The line l crosses the y-axis at C. Find the coordinates of C.

Practice AS examination questions *(continued)*

14 The diagram shows two lines *l* and *m* at right angles to each other. The equation of the line *m* is $3x + 2y = 6$. The line *l* passes through the point with coordinates (5,4).
l and *m* cross the *x*-axis at A and B respectively.

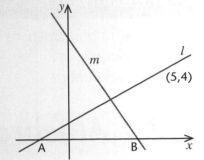

(a) Find the equation of the line *l*.

(b) Show that AB = 3.

15 (a) Find the coordinates of the stationary points on the curve
$y = 2x^3 - 15x^2 - 36x + 10$.

(b) Use the second derivative to determine the nature of the stationary points found in (a).

(c) Find the set of values of *x* for which $2x^3 - 15x^2 - 36x + 10$ is a decreasing function of *x*.

16 The diagram shows a straight line crossing the curve $y = (x - 1)^2 + 4$ at the points P and Q.

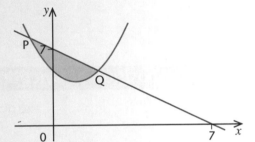

(a) Write down the equation of the straight line through P and Q.

(b) Find the coordinates of P and Q.

(c) Calculate the value of the shaded area.

Chapter 2

Pure 2

The following topics are covered in this chapter:

- Algebra and functions
- Trigonometry
- Differentiation
- Integration
- Numerical methods

2.1 Algebra and functions

After studying this section you should be able to:

- understand composite functions and find the inverse of a one–one function
- sketch the graph of the inverse of a function from the graph of the function
- understand the properties of the modulus function
- use the binomial theorem for positive whole number powers
- carry out algebraic division and understand the terms quotient and remainder
- use the remainder theorem and understand the connection with the factor theorem
- understand the definitions of e^x and $\ln x$
- understand the log laws and use them to solve equations of the form $a^x = b$

LEARNING SUMMARY

Composition of functions

AQA	P1
EDEXCEL	P2
OCR	P2
WJEC	P2
NICCEA	P2

The diagram shows how two functions f and g may be combined.

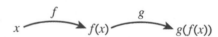

$$x \xrightarrow{\ f\ } f(x) \xrightarrow{\ g\ } g(f(x))$$

In this case, f is applied first to some value x giving $f(x)$. Then g is applied to the value $f(x)$ to give $g(f(x))$. This is usually written as $gf(x)$ and gf can be thought of as a new **composite function** defined from the functions f and g.

For example, if $f(x) = 3x$, $x \in \mathbb{R}$ and $g(x) = x + 2$, $x \in \mathbb{R}$ then
$$gf(x) = g(3x) = 3x + 2.$$

The order in which the functions are applied is important. The composite function fg is found by applying g first and then f.

In this case, $fg(x) = f(x + 2) = 3x + 6$.

Generally speaking, when two functions f and g are defined, the composite functions fg and gf will not be the same.

The inverse of a function

AQA	P1
EDEXCEL	P2
OCR	P2
WJEC	P2
NICCEA	P2

The **inverse of a function** f is a function, usually written as f^{-1}, that *undoes* the effect of f. So the inverse of a function which adds 2 to every value, for example, will be a function that subtracts 2 from every value.

This can be written as $f(x) = x + 2$, $x \in \mathbb{R}$ and $f^{-1}(x) = x - 2$, $x \in \mathbb{R}$.

The domain of f^{-1} is given by the range of f.

Notice that $f^{-1}f(x) = f^{-1}(x + 2) = x$ and that $ff^{-1}(x) = f(x - 2) = x$.

A function can be either one–one or many–one, but only functions that are one–one can have an inverse. The reason is, that reversing a many–one function would give a mapping that is one–many, and this cannot be a function.

The diagram shows a many–one function. It does not have an inverse.

Every value in the domain of a function can have only one image. This is an important property of functions.

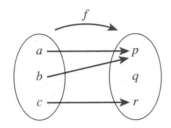

If you reverse this diagram then p would have more than one image.

Some important examples of restricting the domain of a function so that it will have an inverse are found in trigonometry. See page 57.

You can turn a many–one function into a one–one function by restricting its domain.

For example, the function $f(x) = x^2$, $x \in \mathbb{R}$ is many–one and so it does not have an inverse. However, the function $f(x) = x^2$, $x > 0$ is one–one and $f^{-1}(x) = \sqrt{x}$, $x > 0$.

Finding the inverse of a one–one function

AQA   P1
EDEXCEL   P2
OCR   P2
WJEC   P2
NICCEA   P2

One way to find the inverse of a one–one function f is to write $y = f(x)$ and then rearrange this to make x the subject so that $x = f^{-1}(y)$. The inverse function is then usually defined in terms of x to give $f^{-1}(x)$. The domain of f^{-1} is the range of f.

You can turn $f^{-1}(y)$ into $f^{-1}(x)$ by replacing every y with an x.

Example (a) Find the inverse of the function $f(x) = \dfrac{x+2}{x-3}$, $x \neq 3$

(b) Sketch the graphs of $y = f(x)$ and $y = f^{-1}(x)$.

Start by writing $y = f(x)$.

(a) Define $y = \dfrac{x+2}{x-3}$

Re-arrange to find x.

then $y(x-3) = x + 2$

Multiply out the brackets.

$xy - 3y = x + 2$

Collect the x terms on one side of the equation.

$xy - x = 3y + 2$

$x(y-1) = 3y + 2$

$x = \dfrac{3y+2}{y-1}$.

$x = f^{-1}(y)$ so this defines the inverse function in terms of y.

It is usual to express the inverse function in terms of x.

So $f^{-1}(x) = \dfrac{3x+2}{x-1}$, $x \neq 1$.

The denominator cannot be allowed to be zero so 1 is not in the domain of f^{-1}.

The graph of the inverse function is the reflection of the original graph in the line $y = x$. This is true for any inverse function but the same scale must be used on both axes or the effect is distorted.

Another point to notice is that the original graph approaches 1 but doesn't reach it so that 1 does not belong to the range of the function. This ties in with 1 not being in the domain of the inverse function.

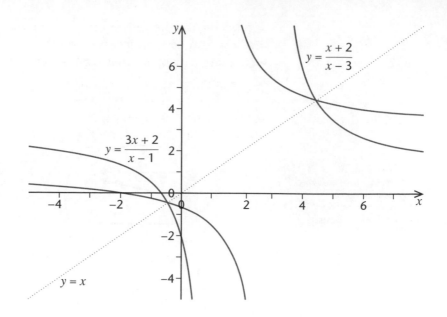

$$y = \frac{x+2}{x-3}$$

$$y = \frac{3x+2}{x-1}$$

$y = x$

The modulus function

AQA — P2
EDEXCEL — P2
OCR — P2
WJEC — P2
NICCEA — P2

The notation $|x|$ is used to stand for the modulus of x. This is defined as

$$|x| = \begin{cases} x & \text{when} \quad x \geqslant 0 \quad \text{(when } x \text{ is positive } |x| \text{ is just the same as } x\text{).} \\ \\ -x & \text{when} \quad x < 0 \quad \text{(when } x \text{ is negative } |x| \text{ is the same as } -x\text{).} \end{cases}$$

The graph of $y = |x|$

$|x|$ is never negative so the graph of $y = |x|$ doesn't go below the x-axis anywhere.

It follows that the graph of $y = |x|$ is the same as the graph of $y = x$ for positive values of x. But, when x is negative, the corresponding part of the graph of $y = x$ must be reflected in the x-axis to give the graph of $y = |x|$.

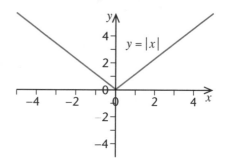

$y = |x|$

The graph of $y = |f(x)|$

The graph of $y = 1/x$ goes below the x axis for negative values of x.

The graph of $y = |f(x)|$ is the same as the graph of $y = f(x)$ for positive values of $f(x)$. But, when $f(x)$ is negative, the corresponding part of the graph of $y = f(x)$ must be reflected in the x-axis to give the graph of $y = |f(x)|$.

The diagram shows the graph of $y = |1/x|$.

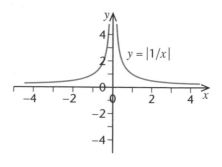

$y = |1/x|$

The graph of $y = f(|x|)$

The graph of $y = f(|x|)$ is the same as the graph of $y = f(x)$ when x is positive. This part of the graph is then reflected in the y-axis to give the graph of $y = f(|x|)$ for negative values of x.

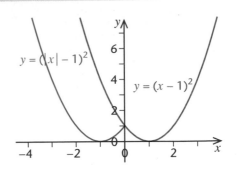

The diagram shows the graphs of $y = (x - 1)^2$ and $y = (|x| - 1)^2$.

The expression $|x - a|$ can be interpreted as the distance between the numbers x and a on the number line. In this way, the statement $|x - a| < b$ means that the distance between x and a is less than b.

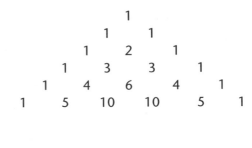

It follows that $a - b < x < a + b$.

Binomial expansion

An expression which has two terms, such as $a + b$ is called a **binomial**. The expansion of something of the form $(a + b)^n$ is called a **binomial expansion**. When n is a positive integer:

> The situation where n is *not* a positive integer can be dealt with but you don't need it for this component.

$$(a + b)^n = a^n + na^{n-1}b + \frac{n(n-1)}{2!} a^{n-2}b^2 + \frac{n(n-1)(n-2)}{3!} a^{n-3}b^3 + \ldots + b^n.$$

This expansion may appear complicated but it starts to look simpler if you follow the patterns from one term to the next:

> Remember that $a^0 = 1$ when $a \neq 0$.

- Starting with a^n, the power of a is reduced by 1 each time until the last term, b^n, which is the same as $a^0 b^n$.

- The power of b is increased by 1 each time. Notice that the first term, a^n, is the same as $a^n b^0$ and that the powers of a and b always add up to n in each term.

- It's worth remembering the first three coefficients: 1, n and $\dfrac{n(n-1)}{2!}$ then the rest follow the same pattern, giving $\dfrac{n(n-1)(n-2)}{3!}$, $\dfrac{n(n-1)(n-2)(n-3)}{4!}$ and so on.

> Pascal's triangle: each row starts and ends with 1. Every other value is found by adding the pair of numbers immediately above it in the pattern.
>
>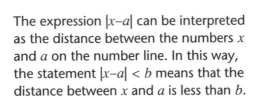

The coefficients in the expansion of $(a + b)^n$ also follow the pattern given by the row of **Pascal's triangle** that starts with 1 n ...
So the coefficients in the expansion of $(a + b)^4$ for example are 1, 4, 6, 4 and 1.

```
            1
          1   1
        1   2   1
      1   3   3   1
    1   4   6   4   1
  1   5  10  10   5   1
```

It follows that
$(a + b)^4 = a^4 + 4a^3 b + 6a^2 b^2 + 4ab^3 + b^4$.

Variations on this result can be obtained by substituting different values for a and b. Some examples are:

> In this case, $a = 1$ and $b = x$.

$$(1 + x)^4 = 1 + 4x + 6x^2 + 4x^3 + x^4.$$

and

> For this one, $a = 1$ and $b = -2x$.

$$(1 - 2x)^4 = 1 + 4(-2x) + 6(-2x)^2 + 4(-2x)^3 + (-2x)^4$$
$$= 1 - 8x + 24x^2 - 32x^3 + 16x^4.$$

The notation $\binom{n}{r}$ is often used to stand for the expression $\dfrac{n(n-1)\ldots(n-r+1)}{r!}$.

So, for example $\binom{n}{3} = \dfrac{n(n-1)(n-2)}{3!}$.

Using this notation, the binomial expansion may be written as:

$$(a+b)^n = a^n + \binom{n}{1}a^{n-1}b + \binom{n}{2}a^{n-2}b^2 + \binom{n}{3}a^{n-3}b^3 + \ldots + b^n.$$

- It's useful to recognise that the term involving b^r takes the form $\binom{n}{r}a^{n-r}b^r$.

- For positive integer values $\binom{n}{r}$ has the same value as ^nC_r and you may find that your calculator will work this out for you.

- When n is small and the full binomial expansion is required, the simplest way to find the coefficients is to use Pascal's triangle. For larger values of n it may be simpler to use the formula, particularly if only some of the terms are required.

Example Expand $(1+3x)^{10}$ in ascending powers of x up to and including the fourth term.

$$(1+3x)^{10} = 1 + \binom{10}{1}(3x) + \binom{10}{2}(3x)^2 + \binom{10}{3}(3x)^3 + \ldots$$

$$= 1 + 10(3x) + 45(9x^2) + 120(27x^3) + \ldots$$

$$= 1 + 30x + 405x^2 + 3240x^3 + \ldots.$$

Example Find the coefficient of the x^7 term in the expansion of $(3-2x)^{15}$

> Notice that the coefficient of x^7 is not totally determined by the value of $\binom{15}{7}$.

The term involving x^7 is given by $\binom{15}{7}(3)^8(-2x)^7$

$$= 6435 \times 6561 \times -128x^7,$$

and so the required coefficient is $-5\,404\,164\,480$.

Algebraic division

AQA   P3
EDEXCEL   P1
OCR   P2
WJEC   P2
NICCEA   P1

If you work out $27 \div 4$ for example, then you obtain 6 as the **quotient** and 3 as the **remainder**. You can write $27 = 4 \times 6 + 3$ to show the connection between these values.

The same thing applies in algebra when one polynomial is divided by another.

Example Find the quotient and remainder when $x^2 + 7x - 5$ is divided by $x - 4$.

> This is an identity. The values of a, b and c may be found by the methods shown in Pure 1. See page 26.

Using $ax + b$ for the quotient and c for the remainder gives:

$$x^2 + 7x - 5 \equiv (x-4)(ax+b) + c. \qquad [1]$$

$$\equiv ax^2 + (b-4a)x + c - 4b.$$

Equating coefficients of x^2 gives: $a = 1$.

Equating coefficients of x gives: $b - 4a = 7$

$$\Rightarrow b - 4 = 7 \qquad \text{(using the result } a = 1 \text{ from above)}$$

$$\Rightarrow b = 11.$$

Equating the constant terms gives: $c - 4b = -5$

$$\Rightarrow c - 44 = -5 \qquad \text{(using the result } b = 11 \text{ from}$$

$$\Rightarrow c = 39. \qquad \text{above)}$$

Substituting for a, b and c in [1] gives $x^2 + 7x - 5 \equiv (x - 4)(x + 11) + 39$.

So, the quotient is $(x + 11)$ and the remainder is 39.

The remainder theorem

AQA	P3
EDEXCEL	P3
OCR	P2
WJEC	P2
NICCEA	P2

> When $x = 4$, $(x - 4) = 0$ so $(x - 4)(ax + b) = 0$ and the RHS simplifies to c.

In the previous example, the quotient and remainder of $(x^2 + 7x - 5) \div (x - 4)$ were found by using $x^2 + 7x - 5 \equiv (x - 4)(ax + b) + c$ and equating coefficients.

Another way to use this identity is to substitute particular values for x. Notice that when $x = 4$, the RHS simplifies to c and so the remainder is easily found by substituting $x = 4$ on the LHS. This gives $c = 4^2 + 7 \times 4 - 5 = 39$ as before.

In its generalised form this result is known as the **remainder theorem** which states:

> **KEY POINT**
>
> When a polynomial $f(x)$ is divided by $(x - a)$ the remainder is $f(a)$.

Example Find the remainder when the polynomial $f(x) = x^3 + x - 5$ is divided by:

(a) $(x - 3)$ (b) $(x + 2)$ (c) $(2x - 1)$.

> Choose x so that $x + 2 = 0$.

> Choose x so that $2x - 1 = 0$.

(a) The remainder is $f(3) = 3^3 + 3 - 5 = 25$.

(b) The remainder is $f(-2) = (-2)^3 - 2 - 5 = -15$.

(c) The remainder is $f(0.5) = 0.5^3 + 0.5 - 5 = -4.375$.

A special case of the remainder theorem occurs when the remainder is 0. This result is known as the **factor theorem** which states:

> **KEY POINT**
>
> If $f(x)$ is a polynomial and $f(a) = 0$ then $(x - a)$ is a factor of $f(x)$.

Example Show that $(x + 3)$ is a factor of $x^3 + 5x^2 + 5x - 3$.

> See page 69 for some other examples of how the factor theorem may be used.

Taking $f(x) = x^3 + 5x^2 + 5x - 3$

$$f(-3) = (-3)^3 + 5(-3)^2 + 5(-3) - 3$$

$$= -27 + 45 - 15 - 3 = 0.$$

By the factor theorem $(x + 3)$ is a factor of $x^3 + 5x^2 + 5x - 3$.

Exponential functions

AQA	P2
EDEXCEL	P2
OCR	P2
WJEC	P2
NICCEA	P2

Exponential functions are often used to represent, or model, patterns of:
- growth when $a > 1$
- decay when $a < 1$.

All scientific calculators will display values of e^x and 10^x. To display the value of e, for example, find e^1.

An **exponential function** is one where the variable is a power or exponent. For example, any function of the form $f(x) = a^x$, where a is a constant, is an exponential function.

The diagram shows some graphs of exponential functions for different values of a.

All of the graphs pass through the point (0,1) and each one has a different gradient at this point. The value of the gradient depends on a.

There is a particular value of a for which the curve has gradient 1 at (0,1). The letter e is used to stand for this value and the function $f(x) = e^x$ is often called **the exponential function** because of its special importance to the subject.
e = 2.718281828 ... it is an irrational number and so it cannot be written exactly.

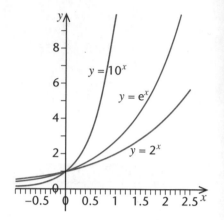

Logarithms

AQA	P2
EDEXCEL	P2
OCR	P2
WJEC	P2
NICCEA	P2

Your calculator will have keys for $\log_{10} x$ and $\ln x$, probably labelled as log and ln.

An exponential function such as $f(x) = 10^x$ is a one–one function and so it has an inverse. If $y = 10^x$ then the inverse function is called the **logarithm of y to base 10**. This function may be used to express x in terms of y. It is written as $x = \log_{10} y$. In the same way, if $y = e^x$ then $x = \log_e y$ but this is usually written as $x = \ln y$. Logarithms to base e are called **natural logarithms** and are needed in integration. Any positive number may be used as the base of a logarithm function, in theory, but the values most commonly used are 10 and e.

For example $10^2 = 100$ and $\log_{10} 100 = 2$.

Try this on your calculator. It helps to show that each function is the inverse of the other.

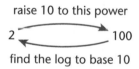

Notice that $\log_a a = 1$ for any base a. In particular $\log_{10} 10 = 1$ and $\ln e = 1$.

Since the exponential and natural logarithm functions are inverses of each other, their graphs are symmetrical about the line $y = x$.

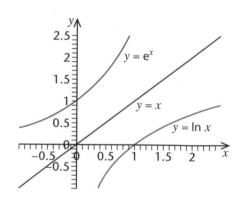

The diagram shows that the domain of the natural logarithm function is given by $x > 0$. This is true for logs to any base.

The log laws

There are three laws of logarithms that you need to know. The same laws apply in any base and so no particular base is stated:

- $\log a + \log b = \log ab$

- $\log a - \log b = \log\left(\dfrac{a}{b}\right)$

- $n \log a = \log a^n$.

You can use these laws to simplify expressions involving logs and to solve **exponential equations**, i.e. equations where the unknown value is a power.

Example Express $\log x + 3 \log y$ as a single logarithm.

$$\log x + 3 \log y = \log x + \log y^3 \qquad \text{(using the third law)}$$
$$= \log xy^3 \qquad \text{(using the first law)}$$

Example Solve the equation $5^x = 30$.

Taking logarithms of both sides gives:

$$\log 5^x = \log 30$$
$$\Rightarrow x \log 5 = \log 30$$

This works in any base.

$$\Rightarrow x = \frac{\log 30}{\log 5} = 2.113 \text{ to 4 s.f.}$$

Progress check

1 $f(x) = x^2 + 5$ and $g(x) = 2x + 3$.

   (a) Write $fg(x)$ in terms of x.

   (b) Find $fg(10)$.

   (c) Find the values of x for which $fg(x) = gf(x)$.

2 Find the inverse of the function $f(x) = \dfrac{x+5}{x-2}$, $x \in \mathbb{R}$, $x \neq 2$ and state the domain of the inverse function.

3 Expand $(1 + 3x)^{12}$ in ascending powers of x up to and including the third term.

4 Find the remainder when $x^3 + 2x^2 - 5x + 1$ is divided by $(x - 2)$.

5 Solve the equation $4^x = 100$ and give your answer to 4 s.f.

5 $x = 3.322$

4 7

3 $1 + 36x + 594x^2$

2 $f(x) = \dfrac{2x+5}{x-1}$, $x \in \mathbb{R}$, $x \neq 1$.

1 (a) $fg(x) = 4x^2 + 12x + 14$ (b) $fg(10) = 534$ (c) $x = -0.08452, -5.915$.

2.2 Trigonometry

After studying this section you should be able to:

- understand the functions secant, cosecant and cotangent and be able to draw their graphs
- understand and use the Pythagorean identities
- understand and use the compound angle formulae for sine, cosine and tangent
- use identities to solve equations

LEARNING SUMMARY

Secant, cosecant and cotangent

AQA	P3
EDEXCEL	P2
OCR	P3
WJEC	P3
NICCEA	P2

These trigonometric functions, commonly known as **sec**, **cosec** and **cot**, are defined from the more familiar sin, cos and tan functions as follows:

$$\sec x = \frac{1}{\cos x} \qquad \operatorname{cosec} x = \frac{1}{\sin x} \qquad \cot x = \frac{1}{\tan x}$$

Each function is periodic and has either line or rotational symmetry. The domain of each one must be restricted to avoid division by zero.

$y = \sec x$

- The period of sec is 360° to match the period of cos.
- Notice that $\sec x$ is undefined whenever $\cos x = 0$.
- The graph is symmetrical about every vertical line passing through a vertex.
- It has rotational symmetry of order 2 about the points in the x-axis corresponding to $90° \pm 180°n$.

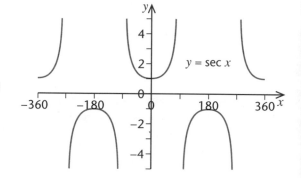

$y = \operatorname{cosec} x$

- The period of cosec is 360° to match the period of sin.
- Notice that $\operatorname{cosec} x$ is undefined whenever $\sin x = 0$.
- The graph is symmetrical about every vertical line passing through a vertex.
- It has rotational symmetry of order 2 about the points on the x-axis corresponding to $0°$, $\pm 180°$, $\pm 360°$, ...

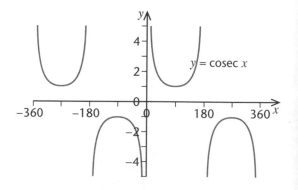

$y = \cot x$

$$\frac{1}{\tan x} = \frac{\cos x}{\sin x}.$$

- The period of cot is 180° to match the period of tan.
- Notice that $\cot x$ is undefined whenever $\sin x = 0$.
- The graph has rotational symmetry of order 2 about the points on the x-axis corresponding to $0°$, $\pm 90°$, $\pm 180°$, ...

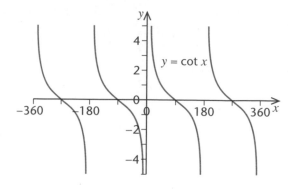

Inverse trigonometric functions

AQA   P3
EDEXCEL   P2
OCR   P3
WJEC   P3
NICCEA   P3

The sine, cosine and tangent functions are all many–one and so do not have inverses on their full domains. However, it is possible to restrict their domains so that each one has an inverse. The graphs of these inverse functions are given below.

> See page 48 to find out about inverse functions.

> You need to know these graphs and to understand the need for restricting the domains.

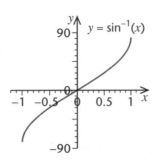
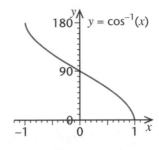
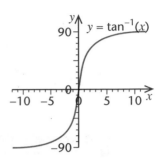

> This is the notation used for inverse trig' functions on your calculator.

> But some texts use:
> arcsin for $\sin^{-1}$
> arccos for $\cos^{-1}$
> arctan for $\tan^{-1}$.

$f(x) = \sin x,\ -90° \leqslant x \leqslant 90°$

$\Rightarrow f^{-1}(x) = \sin^{-1}x,\ -1 \leqslant x \leqslant 1.$

$\sin^{-1}x$ means:
'the angle whose sine is x'
e.g. $\sin^{-1}0.5 = 30°.$

$f(x) = \cos x,\ 0° \leqslant x \leqslant 180°$

$\Rightarrow f^{-1}(x) = \cos^{-1}x,\ -1 \leqslant x \leqslant 1.$

$\cos^{-1}x$ means:
'the angle whose cosine is x'
e.g. $\cos^{-1}0.5 = 60°.$

$f(x) = \tan x,\ -90° < x < 90°$

$\Rightarrow f^{-1}(x) = \tan^{-1}x,\ x \in \mathbb{R}.$

$\tan^{-1}x$ means:
'the angle whose tangent is x'
e.g. $\tan^{-1}1 = 45°.$

Trigonometric identities

AQA   P3
EDEXCEL   P2
OCR   P2
WJEC   P2
NICCEA   P2

If possible, you should *learn* the identities listed below. It is so much easier to simplify expressions, establish new identities and solve equations if you have a firm grasp of these basic results.

> The second result can be obtained from the first by dividing throughout by $\sin^2\theta$.
> The third result can be obtained in a similar way, by dividing throughout by $\cos^2\theta$.

- The Pythagorean identities are:

  $$\sin^2\theta + \cos^2\theta \equiv 1 \qquad 1 + \tan^2\theta \equiv \sec^2\theta \qquad \cot^2\theta + 1 \equiv \operatorname{cosec}^2\theta.$$

- The compound angle formulae for $(A + B)$ are:

  $$\sin(A + B) \equiv \sin A \cos B + \cos A \sin B$$

  $$\cos(A + B) \equiv \cos A \cos B - \sin A \sin B$$

  $$\tan(A + B) \equiv \frac{\tan A + \tan B}{1 - \tan A \tan B}.$$

> These results can be worked out from the ones for $(A + B)$. Just replace every '+' with a '−' and vice versa. This makes it easier to learn the results.

The corresponding results for $(A - B)$ are:

$$\sin(A - B) \equiv \sin A \cos B - \cos A \sin B$$

$$\cos(A - B) \equiv \cos A \cos B + \sin A \sin B$$

$$\tan(A - B) \equiv \frac{\tan A - \tan B}{1 + \tan A \tan B}.$$

You can obtain these results from the compound angle formulae by setting $A = B$.

- The double angle formulae are:

$$\sin 2A \equiv 2 \sin A \cos A$$

$$\cos 2A \equiv \cos^2 A - \sin^2 A$$

$$\equiv 2 \cos^2 A - 1$$

$$\equiv 1 - 2 \sin^2 A$$

These variations are very useful – it is worth knowing them.

$$\tan 2A \equiv \frac{2 \tan A}{1 - 2 \tan^2 A}.$$

- The half angle formulae are:

These results will be useful for integration later in the course.

$$\sin^2 \tfrac{1}{2}A \equiv \tfrac{1}{2}(1 - \cos A)$$

$$\cos^2 \tfrac{1}{2}A \equiv \tfrac{1}{2}(1 + \cos A).$$

Proving identities

AQA P3
EDEXCEL P2
OCR P2
WJEC P3
NICCEA P3

The basic technique for proving a given identity is:

- Start with the expression on one side of the identity.
- Use known identities to replace some part of it with an equivalent form.
- Simplify the result and compare with the expression on the other side.

This is a skill. The more you do, the better you will become.

With practice you get to know which parts to replace so that the required result is established.

Example

Prove that $\dfrac{\cos 2A}{\cos A - \sin A} \equiv \cos A + \sin A.$

Starting with the LHS:

It's often easier to start with the more complicated looking side and try to simplify it.

$$\frac{\cos 2A}{\cos A - \sin A} \equiv \frac{\cos^2 A - \sin^2 A}{\cos A - \sin A}$$

Using one of the double angle results.

$$\equiv \frac{(\cos A + \sin A)(\cos A - \sin A)}{\cos A - \sin A}$$

Using the difference of two squares.

$$\equiv \cos A + \sin A \equiv \text{RHS}.$$

Cancelling the common factor.

It is usually a good idea to replace sec, cosec and cot with sin, cos and tan at the start.

Example

Prove that $\sec^2 A \equiv \dfrac{\text{cosec } A}{\text{cosec } A - \sin A}.$

$$\text{RHS} \equiv \frac{\dfrac{1}{\sin A}}{\dfrac{1}{\sin A} - \sin A} \equiv \frac{1}{1 - \sin^2 A}$$

Replacing cosec A with $\dfrac{1}{\sin A}$ and multiplying the numerator and denominator by $\sin A$.

$$\equiv \frac{1}{\cos^2 A} \equiv \sec^2 A \equiv \text{LHS}.$$

Using one of the Pythagorean identities and the definition of sec A.

Solving equations

AQA ▷ P3
EDEXCEL ▷ P2
OCR ▷ P3
WJEC ▷ P3
NICCEA ▷ P2

Identities may be used to re-write an equation in a form that is easier to solve.

Example Solve the equation $\cos 2x + \sin x + 1 = 0$ for $-\pi \leqslant x \leqslant \pi$.

Replacing $\cos 2x$ with $1 - 2\sin^2 x$ gives $1 - 2\sin^2 x + \sin x + 1 = 0$

$$\Rightarrow 2\sin^2 x - \sin x - 2 = 0.$$

> You need to recognise that this is a quadratic in $\sin x$.
> It has to be rearranged in order to use the quadratic formula.

Using the formula gives $\sin x = \dfrac{1 \pm \sqrt{1+16}}{4} = \dfrac{1 \pm \sqrt{17}}{4}$

$$\Rightarrow \sin x = 1.28 \ldots \text{ (no solutions) or } \sin x = -0.78077 \ldots$$

Using $\sin^{-1}$ gives $\quad x = -0.8959$ to 4 d.p.

> $-1 \leqslant \sin x \leqslant 1$ for all values of x so only one of the roots of the quadratic produces some solutions to the problem.

Another solution in the interval $-\pi \leqslant x \leqslant \pi$ is $-\pi + 0.8959$ giving $x = -2.2457$ to 4 d.p.

The diagram shows there are no other solutions in this interval.

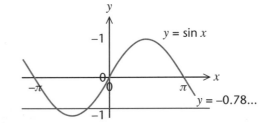

> It's simplest to select a form in which α is acute. But don't worry, exam questions usually specify which form to use.

An important result to be aware of is that the expression $a\cos\theta + b\sin\theta$ can be written in any of the equivalent forms $r\cos(\theta \pm \alpha)$ or $r\sin(\theta \pm \alpha)$ for a suitable choice of r and α. The convention is to take $r > 0$.
One application of the result is to solve equations of the form $a\cos\theta + b\sin\theta = c$.

Example Express $3\cos\theta - 4\sin\theta$ in the form $r\cos(\theta + \alpha)$, $r > 0$. Use the result to solve the equation $3\cos\theta - 4\sin\theta = 2$ for $-\pi \leqslant \theta \leqslant \pi$.

> You need to expand $r\cos(\theta + \alpha)$ and match corresponding parts of the identity.

$$r\cos(\theta + \alpha) \equiv 3\cos\theta - 4\sin\theta$$
$$\Rightarrow r\cos\theta\cos\alpha - r\sin\theta\sin\alpha \equiv 3\cos\theta - 4\sin\theta$$
$$\Rightarrow r\cos\alpha = 3 \text{ and } r\sin\alpha = 4$$

> Notice $\sin\alpha > 0$ and $\cos\alpha > 0$ so α must be acute.
> You could use $\alpha = \sin^{-1}(\frac{4}{5})$.

$$\Rightarrow r = \sqrt{3^2 + 4^2} = 5 \text{ and } \alpha = \cos^{-1}(\tfrac{3}{5}) = 0.9273 \text{ to 4 d.p.}$$
$$\Rightarrow 3\cos\theta - 4\sin\theta \equiv 5\cos(\theta + 0.9273).$$

Using this result in the equation $3\cos\theta - 4\sin\theta = 2$ gives

$$5\cos(\theta + 0.9273) = 2$$
$$\Rightarrow \cos(\theta + 0.9273) = 0.4.$$

This gives $\theta + 0.9273 = \pm 1.1593 \Rightarrow \theta = 0.232$ or $\theta = -2.087$ to 3 d.p.

Progress check

1 Prove the identity $\sin 2\theta \equiv \dfrac{2\tan\theta}{1 + \tan^2\theta}$.

2 Solve the equation $\cos 2x - \sin x = 0$ for $0° \leqslant x \leqslant 360°$.

3 (a) Write $5\sin x + 12\cos x$ in the form $r\sin(x + \alpha)$ where $0 < \alpha < \dfrac{\pi}{2}$.

   (b) Solve the equation $5\sin x + 12\cos x = 7$ for $-\pi \leqslant x \leqslant \pi$.

3 (a) $13\sin(x+1.176)$ (b) $x = -0.607$ or $x = 1.397$ to 3 d.p.

2 $x = 30°, 150°$ or $270°$

1 RHS $= \dfrac{2\tan\theta}{\sec^2\theta} = 2\tan\theta \times \cos^2\theta \equiv 2\sin\theta\cos\theta \equiv \sin 2\theta \equiv$ LHS.

2.3 Differentiation

After studying this section you should be able to:

- *differentiate functions based on e^x and $\ln x$*
- *use the chain rule to differentiate composite functions and to find connected rates of change*

LEARNING SUMMARY

The exponential function e^x

AQA	P2
EDEXCEL	P2
OCR	P2
WJEC	P2
NICCEA	P2

The exponential function was defined on page 54. It has the special property that it is unchanged by differentiation. So, if $y = e^x$ then $\dfrac{dy}{dx} = e^x$

Constant multiples of the exponential function behave in the same way, so if

$y = ae^x$ then $\dfrac{dy}{dx} = ae^x$ where a is any constant.

Examples
Differentiate with respect to x:

(a) $y = 2e^x$ (b) $y = x^3 + e^x$ (c) $f(x) = \dfrac{1}{x} - 3e^x$

(a) $\dfrac{dy}{dx} = 2e^x$ (b) $\dfrac{dy}{dx} = 3x^2 + e^x$ (c) $f'(x) = -\dfrac{1}{x^2} - 3e^x$

> The general result is needed for AQA and OCR.

In general, if $y = ae^{kx}$ then $\dfrac{dy}{dx} = ake^{kx}$

For example, if $y = 3e^{5x}$ then $\dfrac{dy}{dx} = 3 \times 5e^{5x} = 15e^{5x}$.

The natural logarithm function $\ln x$

AQA	P2
EDEXCEL	P2
OCR	P2
WJEC	P2
NICCEA	P2

If $y = \ln x$ then $\dfrac{dy}{dx} = \dfrac{1}{x}$.

This result is easily extended to functions of the form $y = \ln kx$ using the first law of logarithms.

>
> $\log a + \log b = \log ab$
> See page 55.

If $y = \ln kx$ (k is a constant)

then $y = \ln x + \ln k$

giving, $\dfrac{dy}{dx} = \dfrac{1}{x} + 0$ (since $\ln k$ is a constant, its derivative is zero)

so, $\dfrac{dy}{dx} = \dfrac{1}{x}$.

For example, if $y = \ln 5x + 3e^x$ then $\dfrac{dy}{dx} = \dfrac{1}{x} + 3e^x$

The chain rule

AQA	P3
EDEXCEL	P3
OCR	P2
WJEC	P2
NICCEA	P2

The **chain rule** can be written as $\dfrac{dy}{dx} = \dfrac{dy}{du} \times \dfrac{du}{dx}$.

It shows how to differentiate a composite function by separating the process into simpler stages and combining the results.

Example   Differentiate   (a) $y = (3x + 2)^{10}$

(b) $y = \sqrt{x^2 - 9}$.

(a) Taking $u = 3x + 2$ gives $y = u^{10}$ and $\dfrac{dy}{du} = 10u^9$, so $\dfrac{dy}{du} = 10(3x + 2)^9$.

It also gives $\dfrac{du}{dx} = 3$.

So, from the chain rule: $\dfrac{dy}{dx} = 10(3x + 2)^9 \times 3 = 30(3x + 2)^9$.

(b) Taking $u = x^2 - 9$ gives $y = \sqrt{u} = u^{\frac{1}{2}}$, and $\dfrac{dy}{du} = \tfrac{1}{2} u^{-\frac{1}{2}}$, so $\dfrac{dy}{du} = \tfrac{1}{2}(x^2 - 9)^{-\frac{1}{2}}$.

It also gives $\dfrac{du}{dx} = 2x$.

So, from the chain rule: $\dfrac{dy}{dx} = \tfrac{1}{2}(x^2 - 9)^{-\frac{1}{2}} \times 2x = \dfrac{x}{\sqrt{x^2 - 9}}$.

The chain rule can also be used to establish results for **connected rates of change**.

For example, the rate of change of the volume of a sphere can be written as $\dfrac{dv}{dt}$.

The corresponding rate of change of the radius of the sphere can be written as $\dfrac{dr}{dt}$. Using the chain rule, the connection between these rates of change is given by:

$$\frac{dv}{dt} = \frac{dv}{dr} \times \frac{dr}{dt}.$$

Also, since $v = \tfrac{4}{3}\pi r^3$, it follows that $\dfrac{dv}{dr} = 4\pi r^2$, so, the connection can now be written as $\dfrac{dv}{dt} = 4\pi r^2 \dfrac{dr}{dt}$.

Progress check

Find $\dfrac{dy}{dx}$ in each of the following questions.

1 (a) $y = 10e^x$ (b) $y = 4e^x - x^3$ (c) $y = 4e^{2x}$

2 (a) $y = \ln 2x$ (b) $y = \ln 7x - 3e^x$

3 $y = (1 - \sqrt{2}x)^{11}$

2.4 Integration

After studying this section you should be able to:

- integrate $\frac{1}{x}$ and simple variations of this function with respect to x
- integrate e^x and simple variations of this function with respect to x
- use integration to find a volume of revolution

LEARNING SUMMARY

The reciprocal function $\frac{1}{x}$

AQA	P2
EDEXCEL	P2
OCR	P2
WJEC	P2
NICCEA	P2

The result $\int x^n \, dx = \dfrac{x^{n+1}}{n+1} + c$ holds for any value of n other than $n = -1$.

In this special case, the result takes a different form: $\int \dfrac{1}{x} \, dx = \ln|x| + c$.

This extends to $\int \dfrac{k}{x} \, dx = k \int \dfrac{1}{x} \, dx = k \ln|x| + c$,

and to $\int \dfrac{1}{kx} \, dx = \dfrac{1}{k} \int \dfrac{1}{x} \, dx = \dfrac{1}{k} \ln|x| + c$. (where k is a constant)

Remember that $x^{-1} = \dfrac{1}{x}$.

For example, $\int \dfrac{2}{x} \, dx = 2 \ln|x| + c$ and $\int \dfrac{1}{3x} \, dx = \dfrac{1}{3} \ln|x| + c$.

You might need to do some algebraic manipulation first, to put a function in the right form to use these results.

You need to recognise that $\dfrac{x+3}{x}$ is the same as $\dfrac{x}{x} + \dfrac{3}{x} = 1 + \dfrac{3}{x}$.

Example Integrate $\dfrac{x+3}{x}$ with respect to x.

$$\int \dfrac{x+3}{x} \, dx = \int 1 + \dfrac{3}{x} \, dx$$
$$= x + 3 \ln|x| + c.$$

The exponential function e^x

AQA	P2
EDEXCEL	P2
OCR	P2
WJEC	P2
NICCEA	P2

The fact that differentiation leaves the exponential function unchanged was stated on page 60. Since integration is the reverse process of differentiation it follows that

$$\int e^x \, dx = e^x + c \quad \text{and} \quad \int ae^x \, dx = a \int e^x \, dx = ae^x + c.$$

This extension is not required for EDEXCEL.

An extension of this result is $\int e^{kx} \, dx = \dfrac{1}{k} e^{kx} + c$, where k is a constant.

Examples

1 $\int 4e^x \, dx = 4e^x + c$ 2 $\int \dfrac{1}{\sqrt{3}} e^x \, dx = \dfrac{1}{\sqrt{3}} e^x + c$

3 $\int e^{\frac{x}{2}} \, dx = \dfrac{1}{\left(\frac{1}{2}\right)} e^{\frac{x}{2}} + c = 2e^{\frac{x}{2}} + c.$

Volume of revolution

AQA   P3
EDEXCEL   P2
OCR   P2
WJEC   P3
NICCEA   P2

The shaded region R is bounded by the curve $y = f(x)$, the x-axis and the lines $x = a$ and $x = b$.

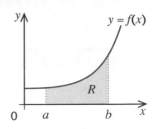

Rotating R completely about the x-axis forms a solid figure. The volume of this figure is called the volume of revolution and is given by

$$V = \int_a^b \pi y^2 \, dx$$

The corresponding result for rotation about the y-axis is

$$\int_a^b \pi x^2 \, dy$$

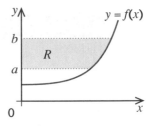

where the region is bounded by the curve $y = f(x)$, the y-axis and the lines $y = a$ and $y = b$.

Example   Find a formula for the volume of a cone of height h and base radius r.

The volume of the cone is given by the volume of revolution of the shaded region shown.

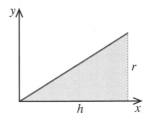

The straight line has gradient $\dfrac{r}{h}$ and passes through the origin, so its equation is $y = \dfrac{r}{h} x$.

Using the formula for volume of revolution gives:

$$V = \int_0^h \pi \left(\frac{r}{h} x\right)^2 dx = \frac{\pi r^2}{h^2} \int_0^h x^2 \, dx$$

$$= \frac{\pi r^2}{h^2} \left[\frac{x^3}{3}\right]_0^h = \frac{\pi r^2}{h^2} \times \frac{h^3}{3}$$

$$= \frac{\pi r^2 h}{3}.$$

Progress check

1. Integrate with respect to x.

   (a) $\dfrac{3}{x}$ (b) $\dfrac{x+3}{x^2}$ (c) $3e^x - \dfrac{1}{x}$.

2. The region bounded by the curve $y = e^{\frac{x}{2}}$, the lines $x = 1$ and $x = 3$ and the x-axis is rotated completely about the x-axis. Find the volume of revolution.

2. $\displaystyle \int_3^1 \pi e^x \, dx = \pi [e^x]_1^3 = \pi (e^3 - e).$

1. (a) $3 \ln|x| + c$ (b) $\ln|x| - \dfrac{3}{x} + c$ (c) $3e^x - \ln|x| + c$

2.5 Numerical methods

After studying this section you should be able to:

- locate an interval containing a root of an equation by finding a change of sign
- use simple iteration to solve equations
- find the approximate value of an integral using the trapezium rule

LEARNING SUMMARY

Change of sign

AQA	P1
EDEXCEL	P2
OCR	P2
WJEC	P2
NICCEA	P2

If there is an interval from $x = a$ to $x = b$ in which the graph of a function has no breaks then the function is said to be **continuous** on the interval.

> All polynomial functions are continuous everywhere.

For example, the graph of $y = x^2 - 3$ is continuous everywhere.

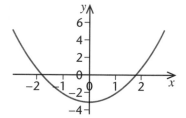

However, the graph of $y = \dfrac{1}{x - 3}$ is not continuous on any interval that contains the value $x = 3$.

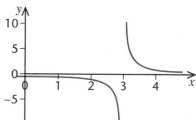

> **KEY POINT**
>
> If $f(x)$ is continuous between $x = a$ and $x = b$ and if $f(a)$ and $f(b)$ have different signs, then a root of $f(x) = 0$ lies in the interval from a to b.

Example   Given that $f(x) = e^x - 10x$, show that the equation $f(x) = 0$ has a root between 3 and 4.

$f(3) = -9.914\ldots < 0$

$f(4) = 14.598\ldots > 0$.

The change of sign shows that $f(x) = 0$ has a root between 3 and 4.

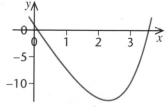

> Most graphic calculators can tabulate values of a function and this gives a very efficient way to obtain the information.

You can continue this process, called a **decimal search**, to trap the root in a smaller and smaller interval. The table gives the values of the function in steps of 0.1.

x	3.1	3.2	3.3	3.4	3.5	3.6	3.7	3.8	3.9
$f(x)$	−8.802	−7.467	−5.887	−4.035	−1.884	0.5982	3.4473	6.7011	10.402

This shows that the root lies between 3.5 and 3.6
At the mid-point $f(3.55) = -0.6866\ldots < 0$ and so the root lies between 3.55 and 3.6.

> This is a good way to establish the value of a root to a particular level of accuracy.

3.5	3.55	3.6

It follows that the value of the root is 3.6 to 1 d.p.

Iteration

AQA	P1
EDEXCEL	P2
OCR	P2
WJEC	P2
NICCEA	P2

To solve an equation by **iteration** you start with some approximation to a root and improve its accuracy by substituting it into a formula. You can then repeat the process until you have the desired level of accuracy.

> There are different methods for producing an iterative formula. You only need to know about simple iteration for this component.

Using **simple iteration**, the first step is to rearrange the equation to express x as a function of itself. This function defines the iterative formula that you need.

For example, the equation $e^x - 10x = 0$ can be rearranged as

$$e^x = 10x$$
$$\Rightarrow x = \ln(10x).$$

This may now be turned into an iterative formula for finding x.

$$x_{n+1} = \ln(10x_n).$$

Starting with $x_1 = 3$, $x_2 = \ln 30 = 3.401 \ldots$ and so on. This produces a sequence of values:

> The values may be found very easily using a calculator's Ans function:
> Key in 3 =
> followed by
> $\ln(10\text{Ans}) = = =$.
> to produce the sequence.
> (Depending on the make of your calculator you may need to use EXE or ENTER in place of = .)

$x_2 = 3.401 \ldots$	$x_6 = 3.5760 \ldots$
$x_3 = 3.526 \ldots$	$x_7 = 3.5768 \ldots$
$x_4 = 3.563 \ldots$	$x_8 = 3.5770 \ldots$
$x_5 = 3.573 \ldots$	$x_9 = 3.5771 \ldots$

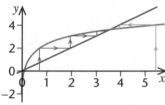

The calculator display soon settles on 3.577152064. The value of the root is $x = 3.577$ to 4 s.f.

The diagram shows how the process converges from a starting point on either side of the root.

> Start from a point on the x-axis, move up to the curve and across to $y = x$. Then move up to the curve and across to $y = x$ again. Continue in the same way. Each of these stages corresponds to one iteration of the formula.

A different rearrangement of the original equation gives $x = \dfrac{e^x}{10}$ and so the corresponding iterative formula is $x_{n+1} = \dfrac{e^{x_n}}{10}$.

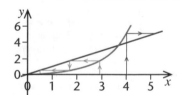

> You may need to try several rearrangements in order to converge to a particular root of an equation by this method.

Starting with $x = 3$, this rearrangement fails to converge to the root between 3 and 4 but it does converge to the other root of the equation.

> Notice how the movement is away from the upper root this time. If the starting value is smaller than the upper root then the process converges to the lower root.
> A starting value above the upper root fails to converge to either root.

The value of this root is $x = 0.1118$ to 4 s.f.

> In some exam questions you may be given the iteration formula to start with.

Example Starting with $x_1 = 1$, use the formula $x_{n+1} = \dfrac{x_n^2 - 5x_n}{10} - 3$ to find $x_2, x_3, \ldots, x_8$ and describe the long term behaviour of the sequence.

> Try the key sequence:
> 1 =
> (Ans² – 5Ans)/10 – 3
> =
> =
> =

$x_2 = -3.4$
$x_3 = -0.144$
$x_4 = -2.9259264$
$x_5 = -0.6809322$
$x_6 = -2.6131669$
$x_7 = -1.0105523$
$x_8 = -2.3926032$

The sequence oscillates but converges to -1.787 to 4 s.f. which is the lower root of the equation

$$x = \frac{x^3 - 5x}{10} - 3,$$

which simplifies to

$$x^2 - 15x - 30 = 0.$$

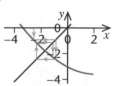

> The oscillations correspond to a cobweb pattern on the diagram.

Numerical integration

AQA	P1
EDEXCEL	P2
OCR	P2
WJEC	P2
NICCEA	P2

The value of $A = \int_a^b f(x)\mathrm{d}x$ represents the area under the graph of $y = f(x)$ between $x = a$ and $x = b$. You can find an approximation to this value using the **trapezium** rule. This is especially useful if the function is difficult to integrate.

The area under the curve between a and b may be divided into n strips of equal width d.

It follows that $d = \dfrac{b-a}{n}$.

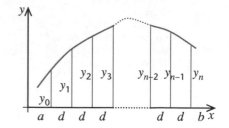

Each strip is approximately a trapezium and so the total area is approximately

$$\tfrac{1}{2}d(y_0 + y_1) + \tfrac{1}{2}d(y_1 + y_2) + \tfrac{1}{2}d(y_2 + y_3) + \ldots + \tfrac{1}{2}d(y_{n-2} + y_{n-1}) + \tfrac{1}{2}d(y_{n-1} + y_n).$$

This simplifies to give the formula known as the trapezium rule:

$$A \approx \tfrac{1}{2}d(y_0 + 2(y_1 + y_2 + y_3 + \ldots y_{n-1}) + y_n).$$

Example

Use the trapezium rule with five strips to estimate the value of $\int_1^2 2^x\,\mathrm{d}x$.

It's useful to tabulate the information:

x	1	1.2	1.4	1.6	1.8	2
2^x	2	2.2974	2.6390	3.0314	3.4822	4
	y_0	y_1	y_2	y_3	y_4	y_5

> Work to a greater level of accuracy in your table than you intend to give in your final answer.

> In this case, the working was done to 5 s.f. and the final answer is given to 3 s.f.

> A graphic calculator with a numerical integration function gives the value of the area as 2.8853901. This shows that no more than 3 s.f. could be justified.

$$d = \frac{2-1}{5} = 0.2$$

So, $\int_1^2 2^x\,\mathrm{d}x \approx 0.1$

$$(2 + 2(2.2974 + 2.6390 + 3.0314 + 3.4822) + 4)$$

$$= 2.89 \quad \text{to} \quad 3 \text{ s.f.}$$

Progress check

1. Show that the equation $x^3 - x - 7 = 0$ has a root between 2 and 3. Use a decimal search to find the value of the root to 2 d.p.

2. Use simple iteration to refine the value of the root found in question 1 and state its value to 4 d.p.

3. Use the trapezium rule with five strips to find the value of $\int_2^3 x \ln x\,\mathrm{d}x$ to 2 d.p.

3 2.31

2 2.0867

1 2.09

Sample questions and model answers

1

The functions f and g are defined by:

$$f(x) = \sqrt{x - 1} \quad x \geqslant 1$$
$$g(x) = 2x + 3 \quad x \in \mathbb{R}$$

(a) Find an expression for $f^{-1}(x)$ and state its domain and range.

(b) Find the value of:

    (i) $fg(7)$

    (ii) $gf(10)$

    (iii) $f^{-1}g^{-1}(15)$

Set $y = f(x)$ and rearrange to make x the subject.

(a) $y = \sqrt{x - 1}$

$\Rightarrow y^2 = x - 1$

$\Rightarrow x = y^2 + 1$

$\Rightarrow f^{-1}(x) = x^2 + 1$

The domain of f^{-1} is given by the range of f.

The domain of f^{-1} is $x \geqslant 0$ and the range of f^{-1} is $x \geqslant 1$.

The range of f^{-1} is given by the domain of f.

The order in which you apply the functions is important.

(b) (i) $fg(7) = f(17) = 4$

    (ii) $gf(10) = g(3) = 9$

    (iii) $f^{-1}g^{-1}(15) = f^{-1}(6) = 37$

2

(a) Solve the inequality $|2x - 3| < 11$

(b) Sketch the graph of $y = |f(x)|$ where $f(x) = (x - 1)(x - 2)(x - 3)$

This first step is important.

(a) $|2x - 3| < 11$

$\Rightarrow -11 < 2x - 3 < 11$

You can simplify the inequality by doing the same thing to each part.

$\Rightarrow -8 < 2x < 14$ *Add 3 to each part.*

$\Rightarrow -4 < x < 7.$ *Divide each part by 2.*

(b)

Parts of the graph of $y = f(x)$ that would be below the x-axis are reflected in the x-axis to give the graph of $y = |f(x)|$

Sample questions and model answers (continued)

3

(a) Find the x coordinates of the turning point of the curve

$$y = x^2 - \ln x.$$

(b) Use the second derivative to determine the nature of the turning point.

(a) $\dfrac{dy}{dx} = 2x - \dfrac{1}{x}$

At a turning point $\dfrac{dy}{dx} = 0$

$\Rightarrow 2x - \dfrac{1}{x} = 0$

$\Rightarrow 2x^2 - 1 = 0$

$\Rightarrow x = \dfrac{1}{\sqrt{2}}$.

> The curve only exists where $x > 0$ so only the positive root is relevant.

(b) $\dfrac{d^2y}{dx^2} = 2 + \dfrac{1}{x^2}$

> In this case, the second derivative is always positive.

When $x = \dfrac{1}{\sqrt{2}}$, $\dfrac{d^2y}{dx^2} = 2.5 > 0$,

so the turning point is a minimum.

4

The diagram shows a region R bounded by the y-axis, the curve $y = x^3$ and the line $y = 8$.

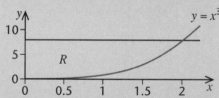

Calculate the volume of the solid figure formed by rotating the region R completely about the y-axis.

For the curve, $y = x^3 \Rightarrow x^2 = y^{\frac{2}{3}}$.

> In the exam, you might be asked to rotate the region about either axis. Make sure that you apply the correct version of the formula.

Volume of revolution $= \displaystyle\int_0^8 \pi x^2\, dy = \int_0^8 \pi y^{\frac{2}{3}}\, dy$

$= \pi \left[\dfrac{3y^{\frac{5}{3}}}{5} \right]_0^8$

$= \dfrac{96\pi}{5}$.

Practice examination questions

1 The functions f and g are defined by:

$$f{:}x \longmapsto 5x - 7 \qquad x \in \mathbb{R}$$

$$g{:}x \longmapsto (x+1)(x-1) \quad x \in \mathbb{R}.$$

(a) Find the range of g.

(b) Find an expression for $fg(x)$.

(c) Determine the values of x for which $fg(x) = f(x)$.

2 Expand $(1 + x)^5$ in ascending powers of x.

Hence find the coefficient of y^6 in the expansion of $(1 - 4y^2)^5$.

3 (a) Given that $(x + 2)$ is a factor of $f(x) = x^3 - 4x^2 - 3x + k$, use the factor theorem to find the value of k.

(b) Factorise $f(x)$ completely and sketch the graph of $y = f(x)$.

(c) Solve the inequality $f(x) \geqslant 0$.

(d) Find the remainder when $f(x)$ is divided by $(x + 1)$.

4 (a) Express $3 \ln x - 2 \ln y + \ln(x + 1)$ as a single logarithm.

(b) Solve the equation $5^x = 100$. Give your answer to 3 d.p.

(c) Solve the equation $\ln(2x + 3) = 4$. Give your answer to 3 d.p.

5 (a) Prove that $\tan x + \cot x \equiv \sec x \operatorname{cosec} x$.

(b) Solve the equation $\cot^2 x + \operatorname{cosec} x = 11$ for values of x in the range $-\pi \leqslant x \leqslant \pi$.

6 (a) Express $15 \sin x + 8 \cos x$ in the form $R \sin(x + \alpha)$ where $0 < \alpha < 90°$.

(b) State the maximum value of $15 \sin x + 8 \cos x$.

(c) Solve the equation $15 \sin x + 8 \cos x = 12$.
Give all the solutions in the interval $0 < x < 360°$.

Practice examination questions (continued)

7 (a) Solve the equation $e^{2x} - 7e^x + 6 = 0$

(b) The temperature of water, in degrees Celsius, in a kettle t minutes after it is switched off is modelled by

$$\theta = 100e^{-0.1t}$$

According to the model:

(i) What is the temperature of the water when the kettle is switched off?

(ii) How long does it take for the temperature of the water to reach 50 °C?

(iii) What happens to the temperature of the water after a long period of time?

(c) Suggest a way in which the model might be improved, assuming that the kettle is in a room where the surrounding temperature is 18 °C.

8 The diagram shows a region R bounded by the curve $y = 3\sqrt{x}$, the x-axis and the lines $x = 1$ and $x = 4$.

(a) Calculate the area of the region R.

(b) Find the volume of revolution when R is rotated through one complete turn about the x-axis. Express your answer in terms of π.

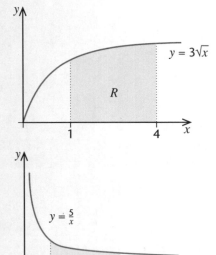

9 (a) Find an expression for the area shown shaded in the diagram in terms of k.

(b) Given that the numerical value of the area is 10, find the value of k to 3 d.p.

10 The population of a country is modelled by $P = Ae^{0.05t}$ where P is the population in millions at time t years and A is a constant.

(a) Show that P is an increasing function of t.

(b) Given that the population when $t = 10$ is 38 million:

(i) Find the value of A and state what this value represents.

(ii) Calculate the size of the population when $t = 20$ to the nearest million.

(c) Find the value of t to the nearest year when the population is expected to reach 50 million.

11 The diagram shows a circle of unit radius.

The angle between the radii is θ where

$0 < \theta < \pi$ and the area of the shaded

segment is $\dfrac{\pi}{5}$.

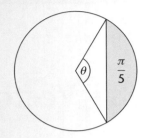

(a) Show that $\theta = \sin\theta + \dfrac{2\pi}{5}$ and use this
result to write an iterative formula for θ.

(b) Find the value of θ to 2 d.p.

12 The area shaded in the diagram is bounded
by the curve $y = \sqrt{x^2 - 3}$, the lines $x = 2$ and
$x = 3$ and the x-axis.

Express the shaded area as an integral and
estimate its value using the trapezium rule
with five intervals. Give your answer to
2 d.p.

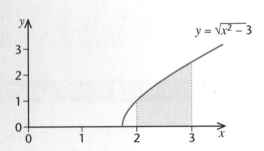

13 (a) Use the chain rule to differentiate:

(i) $y = (4x - 3)^9$

(ii) $y = \ln(5 - 2x)$

(b) Using your results from part (a):

(i) Write down an expression for $\displaystyle\int (4x - 3)^8 \, dx$

(ii) Calculate $\displaystyle\int_3^4 \dfrac{1}{5 - 2x} \, dx$

14 A sphere is increasing in volume at the rate of 200 cm³/s.

Use the chain rule to find the corresponding rate of increase of the radius when
$r = 5$ cm.

Express your answer in terms of π.

Chapter 3
Mechanics 1

The following topics are covered in this chapter:

- *Vectors*
- *Kinematics*

- *Statics*
- *Dynamics*

3.1 Vectors

After studying this section you should be able to:

- *understand the distinction between vector and scalar quantities*
- *add and subtract vector quantities and multiply by a scalar*
- *resolve vector quantities into two perpendicular components*
- *use the unit vectors **i**, **j** and **k***
- *find the magnitude and direction of a vector*

LEARNING SUMMARY

Vector and scalar quantities

AQA	M1
EDEXCEL	M1
OCR	M1
WJEC	M2
NICCEA	M2

A **scalar** quantity has size (or **magnitude**) but not direction. **Numbers** are scalars and some other important examples are **distance, speed, mass** and **time**.

A **vector** quantity has both size and **direction**. For example, distance in a specified direction is called **displacement**. Some other important examples are **velocity, acceleration, force** and **momentum**.

The diagram shows a **directed line segment**. It has size (in this case, length) and direction so it is a vector.

The diagram gives a useful way to represent *any* vector quantity and may also be used to represent addition and subtraction of vectors and multiplication of a vector by a scalar.

> In a textbook, vectors are usually labelled with lower case letters in **bold** print.
> When hand-written, these letters should be underlined e.g. $\underline{a}$

Addition and subtraction of vectors

AQA	M1
EDEXCEL	M1
OCR	M1
WJEC	M2
NICCEA	M2

This diagram shows three vectors **a**, **b** and **c** such that **c** = **a** + **b**.

The vectors **a** and **b** follow on from each other and then **c** joins the start of vector **a** to the end of vector **b**.

> Check that **a** + **b** = **b** + **a**.

The vector **c** is called the **resultant** of **a** and **b**.

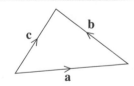

On a vector diagram, –**q** has the opposite sense of direction to **q**.
Notice that **p** – **q** is represented as **p** + (–**q**).

This diagram shows the same information in a different way.

Following the route in the opposite direction to **q** is the same as adding –**q** or subtracting **q**.

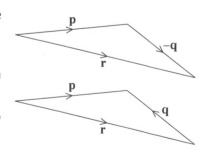

Scalar multiplication

AQA — M1
EDEXCEL — M1
OCR — M1
WJEC — M2
NICCEA — M2

2p is parallel to **p** and has the same sense of direction but is twice as long.

−**3p** is parallel to **p** but has the opposite sense of direction and is three times as long.

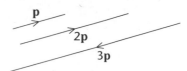

Component form

AQA — M1
EDEXCEL — M1
OCR — M1
WJEC — M2
NICCEA — M2

A unit vector is a vector of magnitude 1 unit.

When working in two dimensions, it is often very useful to express a vector in terms of two special vectors **i** and **j**. These are **unit vectors** at right-angles to each other. A vector **r** written as **r** = a**i** + b**j** is said to have **components** a**i** and b**j**.

Examples

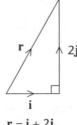

r = **i** + 2**j**

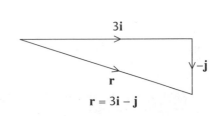

r = 3**i** − **j**

For work involving the Cartesian coordinate system, **i** and **j** are taken to be in the positive directions of the x- and y-axes respectively.

In three dimensions, a third vector **k** is used to represent a unit vector in the positive direction of the z-axis.

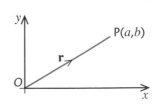

The **position vector** of a point P is the vector $\overrightarrow{OP}$ where O is the origin.

If P has coordinates (a, b) then its position vector is given by **r** = a**i** + b**j**.

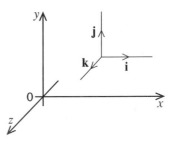

In three dimensions, a point with coordinates (a, b, c) would have position vector **r** = a**i** + b**j** + c**k**.

Resolving a vector

AQA — M1
EDEXCEL — M1
OCR — M1
WJEC — M2
NICCEA — M2

See page 91 for an application of this process.

If you know the magnitude and direction of a vector then you can use trigonometry to **resolve** it in terms of **i** and **j** components.

The magnitude of **r** may be written as | **r** | or simply as r.

From the diagram, **r** = $r \cos \theta$**i** + $r \sin \theta$**j**.

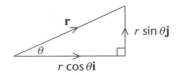

Example

A force **f** has magnitude 10 N at an angle of 60° above the horizontal. Express **f** in the form $a\mathbf{i} + b\mathbf{j}$ where **i** and **j** are unit vectors in the horizontal and vertical directions respectively.

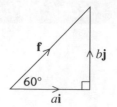

In this case, $a = 10 \cos 60° = 5$ and $b = 10 \sin 60° = 5\sqrt{3}$ giving $\mathbf{f} = 5\mathbf{i} + 5\sqrt{3}\mathbf{j}$.

Adding and subtracting vectors in component form

AQA	M1
EDEXCEL	M1
OCR	M1
WJEC	M2
NICCEA	M2

One advantage of expressing vectors in terms of **i** and **j** is that addition and subtraction may be done algebraically.

Example

$\mathbf{p} = 3\mathbf{i} + 2\mathbf{j}$ and $\mathbf{q} = 5\mathbf{i} - \mathbf{j}$. Express the following vectors in terms of **i** and **j**:

(a) $\mathbf{p} + \mathbf{q}$ (b) $\mathbf{p} - \mathbf{q}$ (c) $2\mathbf{p} - 3\mathbf{q}$.

(a) $\mathbf{p} + \mathbf{q} = (3\mathbf{i} + 2\mathbf{j}) + (5\mathbf{i} - \mathbf{j}) = 8\mathbf{i} + \mathbf{j}$

(b) $\mathbf{p} - \mathbf{q} = (3\mathbf{i} + 2\mathbf{j}) - (5\mathbf{i} - \mathbf{j}) = -2\mathbf{i} + 3\mathbf{j}$

(c) $2\mathbf{p} - 3\mathbf{q} = 2(3\mathbf{i} + 2\mathbf{j}) - 3(5\mathbf{i} - \mathbf{j}) = 6\mathbf{i} + 4\mathbf{j} - 15\mathbf{i} + 3\mathbf{j} = -9\mathbf{i} + 7\mathbf{j}$.

Finding the magnitude and direction of a vector

AQA	M1
EDEXCEL	M1
OCR	M1
WJEC	M2
NICCEA	M2

Another advantage of expressing a vector in terms of **i** and **j** components is that it is easy to find its magnitude and direction.

If $\mathbf{r} = a\mathbf{i} + b\mathbf{j}$ then the magnitude of **r** is given by

$$|\mathbf{r}| = \sqrt{a^2 + b^2}.$$

And the direction of **r** relative to **i** is given by

$$\theta = \tan^{-1}\left(\frac{b}{a}\right).$$

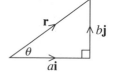

For example, if $\mathbf{r} = 3\mathbf{i} + 4\mathbf{j}$ then $|\mathbf{r}| = \sqrt{3^2 + 4^2} = 5$ and $\theta = \tan^{-1}(\frac{4}{3}) = 53.1°$.

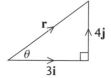

Progress check

1. Write down the position vector **r** of a point with coordinates $(5, -2)$.

2. Given that $|\mathbf{r}| = 12$, express **r** in the form $\mathbf{r} = a\mathbf{i} + b\mathbf{j}$.

3. Find the magnitude of the vector $\mathbf{r} = 5\mathbf{i} - 12\mathbf{j}$ and give its direction relative to **i**.

3 $|\mathbf{r}| = 13$ 67.4° clockwise.
2 $\mathbf{r} = 6\sqrt{3}\mathbf{i} + 6\mathbf{j}$
1 $\mathbf{r} = 5\mathbf{i} - 2\mathbf{j}$

3.2 Kinematics

After studying this section you should be able to:

- *apply the constant acceleration formulae for motion in a straight line*
- *draw and interpret graphs of displacement, velocity and acceleration against time*
- *use calculus for motion with variable acceleration*
- *use vectors to analyse motion in two and three dimensions*
- *apply the constant acceleration formulae to projectiles*

Motion in a straight line

AQA	M1
EDEXCEL	M1
OCR	M1
WJEC	M1
NICCEA	M1

You can use these formulae whenever the acceleration of an object is constant.

$$v = u + at$$

s is the displacement from a fixed position

$$s = ut + \frac{1}{2}at^2$$

u is the initial velocity

$$s = \left(\frac{u+v}{2}\right)t$$

a is the acceleration

t is the time that the object has been in motion

$$v^2 = u^2 + 2as$$

v is the velocity at time t.

Example
An object starts from rest and moves in a straight line with constant acceleration 3 m s^{-2}. Find its velocity after 5 s.

From the given information: $u = 0$

$a = 3$

$t = 5.$

> Don't include units in your working.

Using $v = u + at$

gives $v = 0 + 3 \times 5 = 15.$

> Make a clear statement and include the appropriate units.

The velocity after 5 s is 15 m s^{-1}.

KEY POINT

Motion may take place in either direction along a straight line. One direction is taken to be positive for displacement, velocity and acceleration and the other direction is taken to be negative.

> In some questions the acceleration due to gravity is taken to be 9.8 m s^{-2}.

Example
A stone is thrown vertically upwards with a velocity of 20 m s^{-1}. It has a downward acceleration due to gravity of 10 m s^{-2}.

(a) Find the greatest height reached by the stone.

(b) Find its velocity after 3 s.

(c) Find the height of the stone above the point of projection after 3 s.

(a) From the given information: $u = 20$

$a = -10.$

At the greatest height $v = 0$.
Using $v = u + at$

$0 = 20 - 10t \Rightarrow t = 2.$

Using $s = ut + \frac{1}{2}at^2$

gives $s = 20 \times 2 - 5 \times 4 = 20$

The greatest height reached is 20 m.

(b) Using
$$v = u + at \text{ with } v = 3$$
gives
$$v = 20 - 10 \times 3 = -10.$$
After 3 s the stone is moving downwards at 10 m s^{-1}.

(c) Using
$$s = ut + \tfrac{1}{2}at^2$$
gives
$$s = 20 \times 3 - 5 \times 3^2 = 15.$$

After 3 s the stone is 15 m above the point of projection.

Graphical representation

AQA	M1
EDEXCEL	M1
OCR	M1
WJEC	M1
NICCEA	M1

You need to know the properties of the graphs of distance, displacement, speed, velocity and acceleration against time.

> The gradient of a distance–time graph represents speed.
> The gradient of a displacement–time graph represents velocity.
> The gradient of a velocity–time graph represents acceleration.
>
> The area under a speed–time graph represents the distance travelled.
> The area under a velocity–time graph represents the change of displacement.
> The area under an acceleration–time graph represents change in velocity.

KEY POINT

Example

The diagram represents the progress of a car as it travels between two sets of traffic lights.
Find:

(a) The initial acceleration of the car.

(b) The deceleration of the car as it approaches the second set of lights.

(c) The distance between the traffic lights.

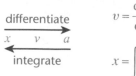

(a) Gradient of OA = $\frac{15}{10} = 1.5$.
The initial acceleration of the car is 1.5 m s^{-2}.

(b) Gradient of BC = $-\frac{15}{15} = -1$.
The *acceleration* of the car is -1 m s^{-2} so the *deceleration* is 1 m s^{-2}.

(c) The area of the trapezium OABC is given by $\frac{15}{2}(5 + 30) = 262.5$,
so, the distance between the traffic lights is 262.5 m.

Variable acceleration

AQA	M2
EDEXCEL	M1
OCR	M1
WJEC	M2
NICCEA	M1

In this case, the constant acceleration formulae do not apply and must not be used.
Using x for displacement, v for velocity and a for acceleration:

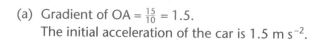

$$v = \frac{dx}{dt} \qquad a = \frac{dv}{dt} = \frac{d^2x}{dt^2}$$

differentiate
$x \quad v \quad a$
integrate

$$x = \int v\,dt \qquad v = \int a\,dt.$$

Example

A particle P moves along the x-axis such that its velocity at time t is given by $v = 5t - t^2$. When $t = 0$, $x = 15$. Find a formula for:

(a) the acceleration of the particle at time t.

(b) the position of the particle at time t.

(a) $v = 5t - t^2 \implies \dfrac{dv}{dt} = 5 - 2t$.

    The acceleration of the particle at time t is given by $a = 5 - 2t$.

(b) $x = \displaystyle\int v \, dt = \int 5t - t^2 \, dt$

    $\implies$ $x = \dfrac{5t^2}{2} - \dfrac{t^3}{3} + c$.

    When $t = 0$, $x = 15$

    so $15 = 0 + c \implies c = 15$.

    This gives the position of the particle at time t as $x = \dfrac{5t^2}{2} - \dfrac{t^3}{3} + 15$.

Using vectors

AQA	M1
EDEXCEL	M1
OCR	M1
WJEC	M2
NICCEA	M2

The work of this section may be extended to two and three dimensions using vectors.

Example

A particle has velocity $2\mathbf{i} - 3\mathbf{j}$ m s^{-1} when $t = 0$ and a constant acceleration of $\mathbf{i} + \mathbf{j}$ m s^{-2}. Find the speed of the particle when $t = 5$ and give its direction.

> The constant acceleration formulae work equally well in two or three dimensions.

From the given information: $\mathbf{u} = 2\mathbf{i} - 3\mathbf{j}$

                               $\mathbf{a} = \mathbf{i} + \mathbf{j}$

                               $t = 5$.

Using $\mathbf{v} = \mathbf{u} + \mathbf{a}t$ gives $\mathbf{v} = 2\mathbf{i} - 3\mathbf{j} + 5(\mathbf{i} + \mathbf{j}) = 7\mathbf{i} + 2\mathbf{j}$.

Speed is the magnitude of the velocity, giving $v = \sqrt{7^2 + 2^2} = 7.28\ldots$.

So the speed of the particle when $t = 5$ is 7.28 m s^{-1} to 2 d.p.

From the diagram, $\tan\theta = \frac{2}{7} \implies \theta = 15.9°$.

When $t = 5$ the particle is moving at $15.9°\ \circlearrowleft$ to the direction of $\mathbf{i}$.

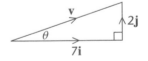

Example

The position vector of a particle at time t is given by $\mathbf{r} = t^3\mathbf{i} + 6t\mathbf{j}$. Find the velocity and acceleration of the particle at time t.

> Differentiate the coefficients of $\mathbf{i}$ and $\mathbf{j}$ with respect to t.

The velocity at time t is given by $\mathbf{v} = \dfrac{d\mathbf{r}}{dt} = \dfrac{d(t^3)}{dt}\mathbf{i} + \dfrac{d(6t)}{dt}\mathbf{j}$

                                     $= 3t^2\mathbf{i} + 6\mathbf{j}$.

The acceleration at time t is given by $\mathbf{a} = \dfrac{d\mathbf{v}}{dt} = 6t\mathbf{i}$.

Projectiles

AQA	M1
EDEXCEL	M2
OCR	M2
WJEC	M2
NICCEA	M2

In the standard model for projectiles, air resistance is ignored and the acceleration due to gravity g m s^{-2} is taken to be constant. According to this model:

> • The only force acting on a projectile, once in flight, is its weight.
> • The horizontal component of velocity is constant.
> • The vertical component of velocity is subject to a constant acceleration g m s^{-2}.

KEY POINT

The approach used for many projectile problems is to resolve the initial velocity into horizontal and vertical components and to treat these two parts separately.

Example

A stone is thrown horizontally with a speed of 10 m s^{-1} from the top of a cliff. The cliff is 98 m high and the stone lands in the sea d m from its base. Take $g = 9.8$ m s^{-2} and show that $d = 20\sqrt{5}$.

> You can assume that a cliff is vertical.

> Choosing downwards to be positive.

> Even a very simple diagram can help you to visualise the problem.

Vertically: $u = 0$

$a = 9.8$

$s = 98$.

Using $s = ut + \frac{1}{2}at^2$

gives $98 = 4.9t^2$

$\Rightarrow t = \sqrt{20} = 2\sqrt{5}$.

Horizontally: $d = 10 \times 2\sqrt{5}$

$\Rightarrow d = 20\sqrt{5}$.

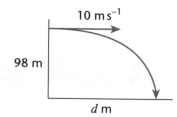

The initial velocity of the projectile may be at angle θ to the horizontal.

Example

A ball is struck with velocity 20 m s^{-1} at 40° to the horizontal from a point 1 m above the ground. Find the maximum height reached by the ball. Take $g = 9.8$ m s^{-2}.

> Choosing upwards to be positive.

Vertically: $u = 20 \sin 40°$

$a = -9.8$.

At max ht: $v = 0$.

Using $v = u + at$

$0 = 20 \sin 40° - 9.8t$

$\Rightarrow t = 1.3118 \dots$.

Using $s = ut + \frac{1}{2}at^2$.

At max ht $s = 20 \sin 40° \times 1.3118 - 4.9 \times 1.3118^2$

$= 8.432 \dots$.

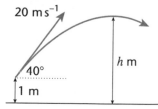

The maximum height is 9.43 m to 2 d.p.

Progress check

1 A stone is thrown vertically upwards with a speed of 15 m s^{-1}. Take the downward acceleration due to gravity to be 10 m s^{-2} and find:
(a) The greatest height reached.
(b) The speed and direction of the stone after 2 s.

2 A particle has position vector $\mathbf{r} = 5t^2\mathbf{i} + t^3\mathbf{j}$ m at time t s.
Find its velocity and acceleration when $t = 3$.

2 $\mathbf{v} = 30\mathbf{i} + 27\mathbf{j}$, $\mathbf{a} = 10\mathbf{i} + 18\mathbf{j}$
1 (a) 33.75 m   (b) 5 m s^{-1} downwards

3.3 Statics

After studying this section you should be able to:

- resolve a single force into perpendicular components
- resolve a system of forces in a given direction
- find the resultant of a system of forces
- solve problems involving friction
- apply the conditions for equilibrium of coplanar forces in simple cases
- find the moment of a force
- locate the position of the centre of mass of a system of particles or a composite body

LEARNING SUMMARY

Force

AQA	M1
EDEXCEL	M1
OCR	M1
WJEC	M1
NICCEA	M1

Force is a vector quantity that influences the motion of an object. It is measured in newtons (N). For example, the **weight** of an object is the force exerted on it by gravity. An object with a mass of m kg has weight mg N. Tension, **reaction** and **friction** are other examples of force and will be met in this section.

Resolving forces

AQA	M1
EDEXCEL	M1
OCR	M1
WJEC	M1
NICCEA	M1

> Notice the directions of the forces shown by the arrows. The diagram represents vector addition of the components.

Here are some examples of resolving forces into two components at right-angles to each other.

In each case the forces are represented in magnitude and direction by the sides of the triangle. The force to be resolved is *always* shown on the hypotenuse.

> **KEY POINT**
>
> These are **vector diagrams** showing the relationship between a force and its components.

> Take care not to show both the force *and* its components on a **force diagram** or you will represent the force TWICE.

In a **force diagram**, you can *replace* a force with a pair of components that are equivalent to it. This will often make it easier to produce the equations necessary to solve a problem.

An important example is that of an object on an **inclined plane**.

The diagram shows an object of weight W on a smooth plane inclined at angle θ to the horizontal. R is the force that the plane exerts on the object. It acts at right-angles to the plane and is called the **normal reaction**.

Force diagram

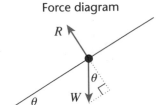

To analyse the behaviour of the object the weight is usually resolved into components parallel and perpendicular to the plane.

> It's worth remembering these results.

This vector diagram shows the component of weight acting down the plane is $W \sin \theta$ and the component perpendicular to the plane is $W \cos \theta$.

Vector diagram

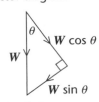

Example

An object of weight 10 N is held in place on a plane inclined at 30° to the horizontal by two forces F N and R N as shown.

Find the values of F and R.

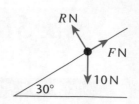

The forces F and R correspond to the components of the weight, parallel and perpendicular to the plane, but act in the opposite directions.

> The forces must balance in each direction.

Resolving parallel to the plane: $F = 10 \sin 30° = 5$.

Resolving perpendicular to the plane: $R = 10 \cos 30° = 5\sqrt{3}$.

Resultant force

AQA	M1
EDEXCEL	M1
OCR	M1
WJEC	M1
NICCEA	M1

The effect of several forces acting at a point is the same as the effect of a single force called the **resultant force**.

You can find the resultant of a set of forces by using vector addition.

Example

Find the magnitude and direction of the resultant $\mathbf{R}$ N of the forces $(3\mathbf{i} - 2\mathbf{j})$ N, $(\mathbf{i} + 3\mathbf{j})$ N and $(-2\mathbf{i} + 4\mathbf{j})$ N.

The resultant force is given by: $\mathbf{R} = (3\mathbf{i} - 2\mathbf{j}) + (\mathbf{i} + 3\mathbf{j}) + (-2\mathbf{i} + 4\mathbf{j})$
$$= 2\mathbf{i} + 5\mathbf{j}.$$

The magnitude of the resultant is $|\mathbf{R}| = \sqrt{2^2 + 5^2} = \sqrt{29}$.

From the diagram: $\theta = \tan^{-1}\left(\frac{5}{2}\right) = 68.2°$.

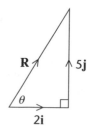

The resultant has magnitude $\sqrt{29}$ N and acts at 68.2° ↺ to the direction of $\mathbf{i}$.

Equilibrium

AQA	M1
EDEXCEL	M1
OCR	M1
WJEC	M1
NICCEA	M1

A set of forces acting at a point is in **equilibrium** if the resultant force is zero.

It follows that the resolved parts of the forces must balance in any chosen direction.

Example

Find the force $\mathbf{F}$ such that the forces $(5\mathbf{i} - 3\mathbf{j})$, $(2\mathbf{i} + \mathbf{j})$ and $\mathbf{F}$ are in equilibrium.

For equilibrium $(5\mathbf{i} - 3\mathbf{j}) + (2\mathbf{i} + \mathbf{j}) + \mathbf{F} = 0$
$$\Rightarrow 7\mathbf{i} - 2\mathbf{j} + \mathbf{F} = 0$$
$$\Rightarrow \mathbf{F} = -7\mathbf{i} + 2\mathbf{j}.$$

Friction

AQA	M1
EDEXCEL	M1
OCR	M1
WJEC	M1
NICCEA	M1

Whenever two rough surfaces are in contact, the tendency of either surface to move relative to the other is opposed by the **force of friction**.

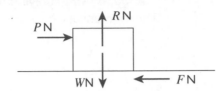

The diagram shows an object of weight W N resting on a rough horizontal surface. The object is pushed from one side by a force P N and friction responds with force F N in the opposite direction.

For small values of P, no movement takes place and $F = P$.

If P increases then F increases to maintain equilibrium until F reaches a maximum value. At this point the object is in **limiting equilibrium** and is on the point of slipping.

If P is now increased again then F will remain the same. The equilibrium will be broken and the object will move.

The maximum value of F depends on:

- The magnitude of the normal reaction R.
- The roughness of the two surfaces measured by the value μ. This value is known as the **coefficient of friction**.

In general: $\qquad\qquad\qquad F \leqslant \mu R$.
For limiting equilibrium: $\qquad F = \mu R$.

Example
An object of mass 8 kg rests on a rough horizontal surface. The coefficient of friction is 0.3 and a horizontal force P N acts on the object which is about to slide. Take $g = 10 \text{ m s}^{-2}$ and find the value of P.

> Remember that the weight of the object is given by mg.

Resolving vertically gives $R = 80$.

Friction is limiting so $F = \mu R = 0.3 \times 80 = 24$.

Resolving horizontally: $P = F \Rightarrow P = 24$.

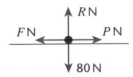

Example
In the diagram, the object of mass m kg is on the point of slipping down the plane. The coefficient of friction between the object and the plane is μ. Show that $\mu = \tan \theta$.

Resolving perpendicular to the plane $\qquad R = mg \cos \theta$.
Resolving parallel to the plane $\qquad\qquad F = mg \sin \theta$.
Friction is limiting so $\qquad\qquad\qquad F = \mu R$.

This gives $\qquad\qquad mg \sin \theta = \mu mg \cos \theta \Rightarrow \mu = \dfrac{\sin \theta}{\cos \theta} = \tan \theta$.

Moments

AQA	M1
EDEXCEL	M1
OCR	M1
WJEC	M1
NICCEA	M1

The **moment of a force** about a point is a measure of the turning effect of the force about the point.

It is found by multiplying the force by its perpendicular distance from the point.

The moment of a force acts either clockwise ↻ or anticlockwise ↺ about the point.

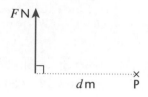

The units of force (N) are multiplied by the units of distance (m) to give N m.

The moment of F about.
P is Fd N m ↺.

The moment of F about P
is $Fd \sin \theta$ N m ↺.

When there is more than one force, the **resultant moment** about a point is found by adding the separate moments. One direction is taken to be positive and the other negative.

Example   Find the resultant moment about O of the forces shown in the diagram.

Taking ↺ as positive:
The total moment is $8 \times 3 - 10 \times 2$ N m ↺
$$= 4 \text{ N m } ↺.$$

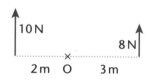

Equilibrium

AQA	M1
EDEXCEL	M1
OCR	M1
WJEC	M1
NICCEA	M1

For an object to be in equilibrium, the resultant force acting on it must be zero and the resultant moment must be zero.

In this context, a light rod is a rod is taken to have a mass that is so small that it can be ignored in any calculations.

Example
The diagram shows a light rod AB in equilibrium. Find the values of F and d.

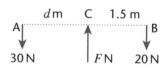

Resolving vertically $F - 30 - 20 = 0$
$\Rightarrow$ $F = 50$

Taking moments about C (written as M(C)) will give an equation involving d.

M(C) gives $30d - 20 \times 1.5 = 0$
$\Rightarrow$ $30d = 30$
$\Rightarrow$ $d = 1.$

Centre of mass

AQA	M1
EDEXCEL	M1
OCR	M2
WJEC	M1
NICCEA	M3

> If m_1 and m_2 are equal then this result gives the mid-point of the two masses.

In the diagram, m_1 and m_2 lie on a straight line through O. Their displacements from O are x_1 and x_2 respectively.

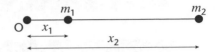

The position given by $\bar{x} = \dfrac{m_1 x_1 + m_2 x_2}{m_1 + m_2}$ is called the **centre of mass** of m_1 and m_2.

In the case where m_1 and m_2 are held together by a light rod, $\bar{x}$ gives the position through which their resultant weight appears to act.
The rod would balance on a support placed in this position.

For n separate masses $m_1, m_2, \cdots m_n$ the position of the centre of mass is given by

$$\bar{x} = \frac{m_1 x_1 + m_2 x_2 + \cdots m_n x_n}{m_1 + m_2 + \cdots m_n}. \text{ This is usually written as } \bar{x} = \frac{\displaystyle\sum_{i=1}^{n} m_i x_i}{\displaystyle\sum_{i=1}^{n} m_i}.$$

Example

In the diagram, the three masses shown lie on a straight line through O. Find the distance of the centre of mass of the system from O.

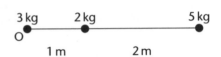

$$\bar{x} = \frac{3 \times 0 + 2 \times 1 + 5 \times 3}{3 + 2 + 5} \text{ m} = 1.7 \text{ m}.$$

The centre of mass is 1.7 m from O.

In two dimensions, using the usual Cartesian co-ordinates, the position of the centre of mass is at $(\bar{x}, \bar{y})$ where $\bar{x}$ and $\bar{y}$ are given by

$$\bar{x} = \frac{\displaystyle\sum_{i=1}^{n} m_i x_i}{\displaystyle\sum_{i=1}^{n} m_i} \text{ as above, and } \bar{y} = \frac{\displaystyle\sum_{i=1}^{n} m_i y_i}{\displaystyle\sum_{i=1}^{n} m_i}.$$

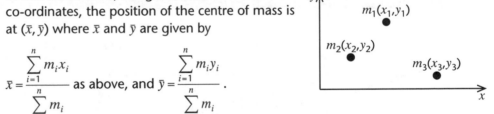

Modelling in mechanics frequently involves such things as **light rigid bodies** that simply act as placeholders for individual masses or **uniform** bodies where the mass is taken to be at the geometric centre. A **lamina** is a thin flat object such as a sheet of metal.

Progress check

1. An object of mass 8 kg rests on a plane inclined at 40° to the horizontal. Find the components of its weight parallel and perpendicular to the plane. Take $g = 9.81$ m s^{-2}.

2. Find the magnitude and direction of the resultant **R** N of the forces $(2\mathbf{i} + 3\mathbf{j})$ N, $(5\mathbf{i} - 4\mathbf{j})$ N and $(-3\mathbf{i} - 7\mathbf{j})$ N.

3. An object of weight 50 N rests on a rough horizontal surface. A horizontal force of 20 N is applied to the object so that it is on the point of slipping. Find the value of the coefficient of friction.

4. The diagram shows a light rod with masses of 5 kg, 2 kg and 1 kg attached. Find the distance of the centre of mass from O.

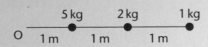

1 parallel: 50.4 N, perp: 60.1 N
2 8.94 N at 63.4° to **i**
3 0.4   4 1.5 m.

3.4 Dynamics

After studying this section you should be able to:

- *understand and apply Newton's laws of motion in two or three dimensions*
- *analyse the motion of connected particles*
- *understand and apply the principle of conservation of momentum*
- *use the relationship between impulse and momentum to solve problems*

LEARNING SUMMARY

Newton's laws of motion

AQA	M1
EDEXCEL	M1
OCR	M1
WJEC	M1
NICCEA	M1

Newton's laws of motion provide us with a clear set of rules that can be used to analyse the effect of forces within a system.

The laws may be stated as:

1. Every particle continues in a state of rest or uniform motion, in a straight line, unless acted upon by an external force.

2. The resultant force acting on a particle is equal to its change of momentum. *This law is most often applied in the form $F = ma$.*

3. Every force has an equal and opposite reaction.

The formula $F = ma$

AQA	M1
EDEXCEL	M1
OCR	M1
WJEC	M1
NICCEA	M1

In the formula $F = ma$, F N stands for the resultant force acting on a particle, m kg is the mass of the particle and a m s^{-2} is the acceleration produced.

It is important to use the correct units.

Example
A particle of mass 2 kg rests on a smooth horizontal plane.
Horizontal forces of 15 N and 4 N act on the particle in opposite directions.
Find the acceleration of the particle.

Using $F = ma$

$$15 - 4 = 2a$$

$$\Rightarrow \qquad a = 5.5.$$

The acceleration of the particle is 5.5 m s^{-2}.

The formula applies equally well in two or three dimensions using vectors.

Example
Forces $(7\mathbf{i} - 2\mathbf{j} + \mathbf{k})$ N and $(3\mathbf{i} + \mathbf{j} - 5\mathbf{k})$ N act on a particle of mass 10 kg.
Find the acceleration produced.

The resultant force is the vector sum of the given forces, so using $\mathbf{F} = m\mathbf{a}$ gives

$$(7\mathbf{i} - 2\mathbf{j} + \mathbf{k}) + (3\mathbf{i} + \mathbf{j} - 5\mathbf{k}) = 10\mathbf{a}$$

$$\Rightarrow \qquad 10\mathbf{i} - \mathbf{j} - 4\mathbf{k} = 10\mathbf{a}$$

$$\Rightarrow \qquad \mathbf{a} = \mathbf{i} - 0.1\mathbf{j} - 0.4\mathbf{k}$$

The acceleration of the particle is $\mathbf{i} - 0.1\mathbf{j} - 0.4\mathbf{k}$ m s^{-2}.

The formula applies even when the force is variable.

Example
A force $(2t\mathbf{i} - 5\mathbf{j})$ N acts on a particle of mass 0.5 kg at time t seconds. Find the velocity of the particle at time t given that, initially, the velocity is $-3\mathbf{i} + 7\mathbf{j}$ m s^{-1}.

Using $\mathbf{F} = m\mathbf{a}$ gives

$$2t\mathbf{i} - 5\mathbf{j} = 0.5\mathbf{a}$$
$$\Rightarrow \qquad \mathbf{a} = 4t\mathbf{i} - 10\mathbf{j}$$
$$\mathbf{v} = \int 4t\mathbf{i} - 10\mathbf{j}\, dt$$
$$\Rightarrow \qquad \mathbf{v} = 2t^2\mathbf{i} - 10t\mathbf{j} + \mathbf{c}.$$

When $t = 0$, $\mathbf{v} = -3\mathbf{i} + 7\mathbf{j} \Rightarrow \mathbf{c} = -3\mathbf{i} + 7\mathbf{j}$.

This gives
$$\mathbf{v} = 2t^2\mathbf{i} - 10t\mathbf{j} - 3\mathbf{i} + 7\mathbf{j}$$
$$\mathbf{v} = (2t^2 - 3)\mathbf{i} + (7 - 10t)\mathbf{j}.$$

The velocity of the particle at time t is $(2t^2 - 3)\mathbf{i} + (7 - 10t)\mathbf{j}$ m s^{-1}.

Connected particles

In a typical problem, two particles are connected by a **light inextensible string**. Since the string is light, there is no need to consider its mass. Since it is inextensible, both particles will have the same speed and accelerate at the same rate *while the string is taut*.

By Newton's third law, the tension in the string acts equally on both particles but in opposite directions.

To solve problems involving connected particles you need to:

• Draw a diagram.

• Identify and label the forces acting on each particle.

• Write the **equation of motion** for each particle i.e. apply $F = ma$ for each one.

• Solve the simultaneous equations produced.

Example
Two particles P and Q are connected by a light inextensible string. The particles are at rest on a smooth horizontal surface and the string is taut. A force of 10 N is applied to particle Q in the direction PQ. P has mass 2 kg and Q has mass 3 kg. Find the tension in the string and the acceleration of the system.

Note the use of the double headed arrow to represent acceleration.

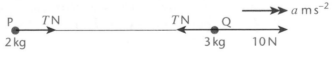

Using $F = ma$

| For particle P | $T = 2a$ | [1] |
| For particle Q | $10 - T = 3a$ | [2] |

[1] + [2] gives $\qquad 10 = 5a \Rightarrow a = 2$
Substituting for a in [1] gives $\qquad T = 4.$

The tension in the string is 4 N and the acceleration of the system is 2 m s^{-2}.

Momentum

AQA	M1
EDEXCEL	M1
OCR	M1
WJEC	M1
NICCEA	M1

The **momentum** of a particle is a vector quantity given by the product of its mass and its velocity i.e. momentum = $m\mathbf{v}$ where m kg is the mass of a particle and $\mathbf{v}$ m s^{-1} is its velocity.

The **principle of conservation of momentum** states that when two particles collide:

the total momentum before impact = the total momentum after impact.

> This kind of diagram gives a useful way to represent the information.

Before impact

After impact

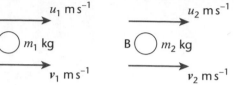

> The positive direction for velocity is shown from left to right.

In the diagram, $u_1 > u_2$ so that the particles collide.

Conservation of momentum gives $m_1 u_1 + m_2 u_2 = m_1 v_1 + m_2 v_2$.

Example

A particle of mass 5 kg moving with speed 4 m s^{-1} hits a particle of mass 2 kg moving in the opposite direction with speed 3 m s^{-1}. After the impact the two particles move together with the same speed v m s^{-1}. Find the value of v.

> Unknown velocities are shown in the positive direction.

Before impact

After impact

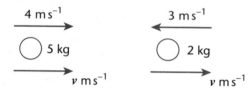

Conservation of momentum gives $5 \times 4 - 2 \times 3 = 5v + 2v$

$\Rightarrow \qquad 7v = 14$

$\Rightarrow \qquad v = 2.$

Impulse

EDEXCEL	M1
OCR	M2
WJEC	M1
NICCEA	M1

The **impulse** of a constant force acting over a given time is given by the product of force and time. This may be written as $I = Ft$ where the impulse is I N s, the force is F N and the time is t s.

It follows that: Impulse = change in momentum

so $\qquad Ft = mv - mu$.

> The total momentum of the system is conserved.

When two particles collide, each receives an impulse from the other of equal size but opposite sign. In this way, the total change in momentum is zero as expected.

Example

A particle of mass 4 kg, initially at rest, is acted upon by a force of 3 N for 10 s. What is the speed of the particle at the end of this time?

Using $\quad Ft = mv - mu$

gives $\quad 3 \times 10 = 4v - 0$

$\Rightarrow \qquad v = 7.5.$

The speed of the particle is 7.5 m s^{-1}.

When two particles collide, the contact force between them may only last a very short time and is unlikely to be constant. However, the value of F in the equation $Ft = mv - mu$ may be used to represent the average value of this force.

Example

A ball of mass 0.5 kg strikes a hard floor with speed 2 m s^{-1} and rebounds with speed 1.5 m s^{-1}.

Given that the ball is in contact with the floor for 0.05 s find the average value of the contact force.

Taking upwards as the positive direction:

$u = 2$, $v = -1.5$, $m = 0.5$ and $t = 0.05$.

Using $Ft = mv - mu$

gives $0.05F = 0.5 \times 1.5 + 0.5 \times 2$

$\Rightarrow \quad 0.05F = 1.75$.

$\Rightarrow \quad F = 35$.

The average value of the contact force is 35 N.

Before impact   After impact

0.5 kg

2 m s^{-1} 1.5 m s^{-1}

Floor

Progress check

1 A particle of mass 8 kg is acted upon by the forces $(11\mathbf{i} - 3\mathbf{j})$ N, $(14\mathbf{i} + \mathbf{j})$ N and $(-\mathbf{i} + 10\mathbf{j})$ N. Find the acceleration of the particle.

2 A particle moves along the x-axis such that its displacement x m from O at time t s is given by $x = t^3 - 15$ for $0 \leqslant t \leqslant 5$.
   (a) Find the acceleration of the particle at time t s.
   (b) Given that the particle has mass 5 kg find the resultant force acting on it when $t = 4$.

3 Two particles A and B are connected by a light inextensible string. The particles are at rest on a smooth horizontal surface and the string is taut. A force of 12 N is applied to particle B in the direction AB. A has mass 4 kg and B has mass 2 kg. Find the tension in the string and the acceleration of the system.

4 A particle of mass 8 kg moving with speed 2 m s^{-1} hits a stationary particle of mass 2 kg.
   After the impact both particles move in the same direction with speed v m s^{-1}.
   (a) Find the value of v.
   (b) Find the impulse given to the stationary particle.

4 (a) $v = 1.6$ (b) Impulse = 3.2 N s.
3 Tension = 8 N, acceleration = 2 m s^{-2}.
2 (a) $6t$ m s^{-2} (b) 120 N.
1 $3\mathbf{i} + \mathbf{j}$ m s^{-2}.

Sample questions and model answers

1

A particle moving in a straight line passes through a fixed point O when $t = 0$. Its velocity at time t seconds is given by $v = 12 - 3t^2 \text{ m s}^{-1}$.

(a) Find the acceleration of the particle when $t = 4$.

(b) Find the distance of the particle from O:

    (i)   when it comes to rest
    (ii)   when $t = 5$.

(a) $v = 12 - 3t^2 \implies a = \dfrac{dv}{dt} = -6t$

     $t = 4 \implies a = -6 \times 4 = -24.$

The acceleration of the particle when $t = 4$ is -24 m s^{-2}.

(b) (i) The particle comes to rest when $v = 0$.

        $v = 0 \implies 12 - 3t^2 = 0$

               $\implies t = 2$ since $t > 0$.

      The displacement of the particle from O at time t is given by

$$x = \int v \, dt + c = \int 12 - 3t^2 \, dt + c$$

      $\implies x = 12t - t^3 + c.$

      When $t = 0$, $x = 0$

      giving $0 = 0 + c \implies c = 0.$

      So $x = 12t - t^3.$

      When $t = 2$ $x = 12 \times 2 - 2^3 = 16.$

      The particle is 16 m from O when it comes to rest.

(b) (ii) When $t = 5$ $x = 12 \times 5 - 5^3 = -65.$

       The particle is 65 m from O when $t = 5$.

Distance is taken to be positive.

Sample questions and model answers (continued)

2

Two particles are connected by a light inextensible string passing over a light frictionless pulley. Particle A has mass 5 kg and lies on a smooth horizontal table. Particle B has mass 4.5 kg and hangs freely. Take $g = 10 \text{ m s}^{-2}$.

(a) Write down the equation of motion for each particle.

(b) Find the acceleration of the system.

(c) Find the tension in the string.

(a)

A clearly labelled diagram is an essential first step.

Both particles have the same acceleration.

The tension is the same at both ends of the string.

The diagram shows the forces acting on each particle in the direction of motion.

Use $F = ma$ for each particle in turn.

(a) The equations of motion are:

for particle A $\qquad\qquad T = 5a$ $\qquad\qquad$ [1]

for particle B $\qquad\qquad 45 - T = 4.5a$ $\qquad\qquad$ [2]

Solve the simultaneous equations and interpret the results.

(b) [1] + [2] gives $\qquad\qquad 45 = 9.5a$

$\qquad\qquad\qquad \Rightarrow a = 4.74$ to 3 s.f.

The acceleration of the system is 4.74 ms^{-2} to 3 s.f.

(c) Substitution for a in [1] gives $T = 23.7$ to 3 s.f.

It is a good idea to state the degree of accuracy given in your answer.

The tension in the string is 23.7 N to 3 s.f.

Practice examination questions

1 A particle moving in a straight line passes through a point O with velocity 10 m s^{-1} when $t = 0$. Given that the acceleration of the particle is -4 m s^{-2}.

Find:

(a) The velocity of the particle when $t = 5$.

(b) The distance that the particle travels between $t = 0$ and $t = 4$.

2

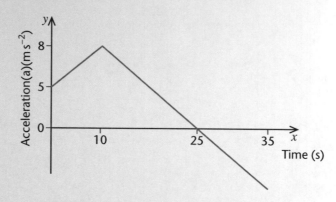

The diagram is an acceleration–time graph for the motion of a particle during a period of 35 s. The particle moves in a straight line and its initial velocity is 20 m s^{-1}.

(a) State the time at which the velocity reaches its maximum value.

(b) Calculate the maximum value of the velocity of the particle during this period.

3 A particle passes through a point P with position vector $(-5\mathbf{i} + 11\mathbf{j} + \mathbf{k})$ with velocity $(2\mathbf{i} + \mathbf{j})$ m s^{-1} when $t = 0$. The particle moves with constant acceleration $(\mathbf{i} + 3\mathbf{j})$ m s^{-2} and reaches the point Q when $t = 4$.

(a) Find $\overrightarrow{PQ}$.

(b) Find the position vector of Q.

(c) Find the average speed of the particle between P and Q.

Practice examination questions (continued)

4   A particle is projected from a point on horizontal ground with speed v at angle θ to the horizontal.

   (a) Find an expression for the maximum height reached during the subsequent motion.

   (b) Find an expression for the horizontal distance travelled before it hits the ground.

5   A ball is kicked with speed 15 m s^{-1} at an angle of 30° to the horizontal from a point on level ground. The ball just clears a fence 2 m high.
Find the maximum possible distance between the fence and the point where the ball was kicked. Take $g = 9.8$ m s^{-2}.

6
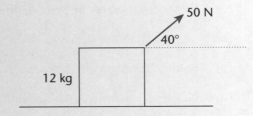

The diagram shows an object of mass 12 kg resting on a rough horizontal surface.

A force of 50 N acts on the object at an angle of 40^0 to the horizontal.
Take $g = 10$ m s^{-2}.

   (a) Find the normal reaction between the horizontal surface and the object.

   (b) Given that the object is in limiting equilibrium, find the coefficient of friction between the object and the horizontal surface.

7   The diagram shows a mass of 6 kg acted upon by a horizontal force of P N.

The coefficient of friction between the object and the plane is 0.3.

Find the minimum value of P required to maintain the object in equilibrium.

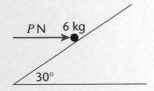

Practice examination questions *(continued)*

8

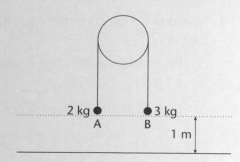

Particles A and B of mass 2 kg and 3 kg respectively are connected by a light inextensible string passing over a smooth frictionless pulley. The system is released from rest with both particles 1 m above the ground. Find:

(a) The acceleration of the system while the string remains taut.

(b) The tension in the string.

(c) The time taken for particle B to reach the ground.

(d) The speed with which particle B hits the ground.

(e) The greatest height reached by particle A.

(f) What difference would it make to your initial equations if the pulley was not light and frictionless?

9

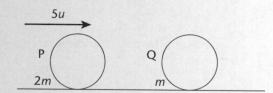

A particle P of mass $2m$ moving with speed $5u$ strikes particle Q of mass m, which is at rest.

After the collision both particles move in the same direction but the speed of particle Q is twice the speed of particle P.

(a) Find the speed of particle P after the collision.

(b) Find the magnitude of the impulse exerted on Q by P during the impact.

Practice examination questions (continued)

10

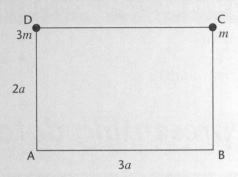

A uniform rectangular lamina ABCD measuring $2a$ by $3a$ has mass $2m$.
Particles of mass m and $3m$ are attached at C and D respectively.

(a) Find the horizontal distance of the centre of mass of the system from AD.

(b) Find the vertical distance of the centre of mass of the system from AB.

(c) The lamina is suspended freely from A. Find the angle that AD makes with the horizontal.

Chapter 4

Statistics 1

The following topics are covered in this chapter:

- Representing data
- Probability
- Discrete random variables

- The Normal distribution
- Correlation and regression

4.1 Representing data

After studying this section you should be able to:

- calculate averages for discrete and continuous data
- find the variance, standard deviation, range and inter-quartile range of data
- draw diagrams to represent and compare distributions
- interpret measures of location and dispersion in comparing sets of data
- understand the concepts of outliers and skewness

LEARNING SUMMARY

Measures of central location

AQA S1
EDEXCEL S1
OCR S1
WJEC S1
NICCEA S1

An average is a value that is taken to be representative of a data set. The three forms of average that you need are the mean, median and mode. These are often referred to as **measures of central location**.

The **mean** $\bar{x}$ of the values $x_1, x_2, x_3, ..., x_n$ is given by $\bar{x} = \dfrac{\sum x_i}{n}$. If each x_i occurs with frequency f_i then the mean of the **frequency distribution** is given by $\bar{x} = \dfrac{\sum f_i x_i}{\sum f_i}$.

Example

Find the mean of these results obtained by throwing a dice.

score	1	2	3	4	5	6
frequency	18	17	23	20	24	18

$$\bar{x} = \frac{18 \times 1 + 17 \times 2 + 23 \times 3 + 20 \times 4 + 24 \times 5 + 18 \times 6}{18 + 17 + 23 + 20 + 24 + 18} = \frac{429}{120} = 3.575$$

The **median** is the middle value of the data when it is arranged in order of size. If there are an even number of values then the median is the mean of the middle pair. In the example above, the 60th and 61st values are both 4 so the median is 4.

The **mode** is the value that occurs with the highest frequency. In the example above the mode is 5.

You can *estimate* the mean of **grouped data** by using the mid-point of each class interval to represent the class.

Example

Estimate the mean value of h from the figures given in the table.
An estimate of the mean is given by

$$\bar{x} = \frac{755}{72} = 10.486 \ldots$$

$10.5 =$ to 1 d.p.

Once data has been grouped, the exact values are not available and so it is only possible to *estimate* the mean.

Interval	frequency (f_i)	midpoint (x_i)	$f_i \times x_i$
$0 < h \leqslant 5$	8	2.5	20
$5 < h \leqslant 10$	24	7.5	180
$10 < h \leqslant 15$	29	12.5	362.5
$15 < h \leqslant 20$	11	17.5	192.5
Totals	72		755

Measures of dispersion

AQA	S1
EDEXCEL	S1
OCR	S1
WJEC	S1
NICCEA	S1

An average alone gives no indication of how widely dispersed the data values are. The simplest measure of dispersion, or spread, is the **range**.

The range of a data set is the difference between largest value and the smallest value.

A slightly more refined approach is to measure the spread of the 'middle half' of the data. The data is put into order and the values Q1, Q2 and Q3 are found that divide the data into quarters.

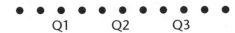

Q1   Q2   Q3

> *The diagram illustrates the situation for a data set of 11 values.*

Q2 is the median of the data.

Q1 is the median of the lower half of the data.

Q3 is the median of the upper half of the data.

The **inter-quartile range** is then Q3 – Q1.

The quartiles may also be used to indicate whether the data values show **positive skew** (Q2 – Q1 < Q3 – Q2) or **negative skew** (Q2 – Q1 > Q3 – Q2).

The **variance** of a data set is the mean of the squares of the deviations from the mean.

> *The second form is easier to work with from raw data.*

$$\text{Variance} = \frac{\sum (x_i - \bar{x})^2}{n} = \frac{\sum x_i^2}{n} - \bar{x}^2.$$

The variance gives an indication of the spread of data values about the mean but the units of the data have been squared in the process.

> *The value may be obtained directly from a scientific calculator but you may need to apply the formula using summary data. See page 106.*

$$\text{Standard deviation} = \sqrt{\text{variance}} = \sqrt{\frac{\sum x_i^2}{n} - \bar{x}^2}.$$

Standard deviation is measured in the same units as the data and is the value most commonly used to measure dispersion at this level.

Statistical diagrams

AQA	S1
EDEXCEL	S1
OCR	S1
WJEC	S1
NICCEA	S1

The use of statistical diagrams may help in comparing distributions and revealing further information about the data.

A **box and whisker plot** for example shows the location and spread of a distribution at a glance.

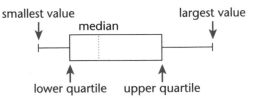

> *Stem and leaf diagrams have the advantage over bar charts that details of the data are available.*

A **back-to-back stem and leaf diagram** allows direct comparison of two sets of data to be made.

The diagram gives a sense of location and spread for each data set.

boys		coursework marks		girls
	2	4	1 7	
	5 3	3	2 5 8 8	
5 4 4 2 1		2	4 4 7 9	
8 7 6 4		1	3 6 6	
6 5		0	7	

2 | 4 means 42%   4 | 1 means 41%

The spread of marks is similar for the boys and the girls but the average for the girls is higher than for the boys.

4.2 Probability

After studying this section you should be able to:

- use Venn diagrams to represent and interpret combined events
- use set notation to express and apply probability laws
- use tree diagrams to represent a problem
- know the formulae for permutation and combinations and recognise when to apply them

LEARNING SUMMARY

Venn diagrams

AQA	S1
EDEXCEL	S1
OCR	S1
WJEC	S1
NICCEA	S1

Venn diagrams provide a useful way to represent information about **sets** of objects.

The reason for mentioning them here is that the diagrams, and the notation used to express results, have a direct interpretation and application in probability theory.

In each diagram, the rectangle represents the set S of all objects under consideration.

A ∪ B is read as A union B.

The circles represent particular sets A and B of objects within the set S.

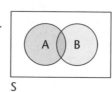

A ∪ B represents the set of all objects that belong to *either* A or B or both.

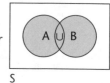

A ∩ B is read as A intersection B.

A ∩ B represents the set of those objects that belong to *both* A *and* B.

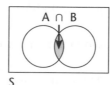

A′ represents the set of objects in S that do not belong to A.

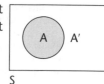

To make use of these ideas in probability theory:

- The objects are interpreted as **outcomes** for a particular situation.
- The set S is the **sample space** of all possible outcomes for the situation.
- The sets A and B are **events** defined by a particular choice of outcomes.
- P(A) for **example**, represents the probability that the event A occurs.
- P(A ∪ B) represents the probability that *either* A *or* B *or both* occurs.
- P(A ∩ B) represents the probability that *both* A *and* B occur.
- P(A′) represents the probability that A does *not* occur.

The **addition rule** may now be written as:

P(A ∪ B) = P(A) + P(B) − P(A ∩ B).

A and B are described as **mutually exclusive events** if they have no outcomes in common. These are events that cannot both occur at the same time.

In this case P(A ∩ B) = 0 and the addition law becomes

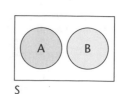

$$P(A \cup B) = P(A) + P(B).$$

P(A ∩ B) = 0

The events A and A′ are mutually exclusive for any event A.
It follows that P(A ∪ A′) = P(A) + P(A′).
Since one of the events A or A′ must occur, P(A ∪ A′) = 1 so P(A) + P(A′) = 1.

This is usually written as $P(A') = 1 - P(A)$.

The **multiplication rule** states that:

$$P(A \cap B) = P(A \mid B) \times P(B),$$

where $P(A \mid B)$ means the probability that A occurs *given that* B has *already* occurred.

$P(A \mid B)$ is described as a **conditional probability**, i.e. it represents the probability of A *conditional* upon B having occurred. Re-arranging the multiplication rule gives

$$P(A \mid B) = \frac{P(A \cap B)}{P(B)}.$$

If A and B are **independent events** then the probability that either event occurs is not affected by whether the other event has already occurred.

In this case, $P(A \mid B) = P(A)$ and the multiplication rule becomes

$$P(A \cap B) = P(A) \times P(B).$$

Example
The events A and B are independent. $P(A) = 0.3$ and $P(B) = 0.6$.
Find (a) $P(A \cup B)$ (b) $P(A' \cap B)$.

(a) (i) $P(A \cup B) = P(A) + P(B) - P(A \cap B)$.

Since A and B are independent, $P(A \cap B) = P(A) \times P(B)$.

> In some problems you need to define the events first before you can apply the probability rules.

So $P(A \cup B) = 0.3 + 0.6 - 0.3 \times 0.6 = 0.72$.

(b) A and B are independent $\Rightarrow$ A' and B are independent.

So $P(A' \cap B) = P(A') \times P(B) = (1 - 0.3) \times 0.6 = 0.42$.

> This is true whether or not the events are independent.

A **tree diagram** is a useful way to represent the probabilities of combined events. Each path through the diagram corresponds to a particular sequence of events and the multiplication rule is used to find its probability. When more than one path satisfies the conditions of a problem, these probabilities are added.

Arrangements and permutations, selections and combinations

EDEXCEL   S1
OCR   S1

There are many situations in probability that can be represented by an arrangement of objects (often letters) in a particular order. The following results are useful for calculating the number of possible arrangements.

Description	No of arrangements
n different objects in a straight line.	$n!$
n objects in a straight line, where r objects are the same and the rest are different.	$\dfrac{n!}{r!}$
n objects in a straight line, where r objects of one kind are the same, q objects of another kind are the same and the rest are different.	$\dfrac{n!}{q!r!}$

The number of arrangements that can be made by choosing r objects from n is $^nP_r = \dfrac{n!}{(n-r)!}$. These are known as **permutations**.

The number of **selections** (the order makes no difference) that can be made by choosing r objects from n is $^nC_r = \dfrac{n!}{(n-r)!r!}$. These are known as **combinations**.

Progress check

1 The probability that the event B occurs is 0.7. The probability that events A and B both occur is 0.4. What is the probability that A occurs given that B has already occurred?

2 Find the number of arrangements of the letters STATISTICS.

3 Find the number of combinations of seven objects chosen from ten.

3 120
2 10! ÷ (3! × 3! × 2!) = 50 400
1 4/7

4.3 Discrete random variables

After studying this section you should be able to:

- understand the concept of a discrete random variable, its expectation and variance
- calculate the expectation and variance of functions of a random variable
- understand the binomial, Poisson and geometric distributions and select the appropriate model for a particular situation

LEARNING SUMMARY

Discrete random variables

AQA	S1
EDEXCEL	S1
OCR	S1
WJEC	S1
NICCEA	S1

The score obtained when a dice is thrown may be thought of as a **random variable**. Since only specific values may be obtained, this is an example of a **discrete random variable**. In the usual notation:

- A capital letter such as X is used as a label for the random variable.
- A lower case letter such as x is used to represent a particular value of X.
- The probability that X takes the value x is written as $P(X = x)$ or just $p(x)$ and this is known as the **probability function**.

> This is a useful result.

- If X is a discrete random variable with probability function $p(x)$ then $\sum p(x) = 1$.

The **probability distribution** of X is given by the set of all possible values of x together with the values of $p(x)$. This is usually shown in a table.

For example, the probability distribution of the scores shown on a dice may be given as:

x:	1	2	3	4	5	6
$p(x)$:	$\frac{1}{6}$	$\frac{1}{6}$	$\frac{1}{6}$	$\frac{1}{6}$	$\frac{1}{6}$	$\frac{1}{6}$

The **cumulative distribution function** is given by $F(x_0) = P(X \leqslant x_0) = \displaystyle\sum_{x \leqslant x_0} p(x)$.

This represents the probability that the random variable X takes a value less than or equal to x_o and is given by the sum of the probabilities up to and including that point.

Expected values

AQA	S1
EDEXCEL	S1
OCR	S1
WJEC	S1
NICCEA	S1

The symbol μ is used to stand for the **mean value** of X. This is also known as the **expected value** of X which is written as $E(X)$.
For any discrete random variable X the mean is $\mu = E(X) = \sum x p(x)$.

Example Find the expected value of the score shown on a dice.

x:	1	2	3	4	5	6
$p(x)$:	$\frac{1}{6}$	$\frac{1}{6}$	$\frac{1}{6}$	$\frac{1}{6}$	$\frac{1}{6}$	$\frac{1}{6}$
$xp(x)$:	$\frac{1}{6}$	$\frac{2}{6}$	$\frac{3}{6}$	$\frac{4}{6}$	$\frac{5}{6}$	$\frac{6}{6}$

$$\mu = \tfrac{1}{6} + \tfrac{2}{6} + \tfrac{3}{6} + \tfrac{4}{6} + \tfrac{5}{6} + \tfrac{6}{6} = \tfrac{21}{6}$$

The expected value is 3.5.

The expected value of a function of a random variable is found in a similar way. In general, if $f(X)$ is some function of the random variable X then

$$E(f(X)) = \Sigma\, f(x)p(x).$$

For example, $E(X^2) = \Sigma\, x^2 p(x)$. In the case of the dice scores above this gives:

$$E(X^2) = \tfrac{1}{6} + \tfrac{4}{6} + \tfrac{9}{6} + \tfrac{16}{6} + \tfrac{25}{6} + \tfrac{36}{6} = \tfrac{91}{6}.$$

Using σ to stand for the standard deviation of X and Var(X) to stand for the variance of X we have:

$$\sigma^2 = \mathrm{Var}(X) = E(X^2) - \mu^2.$$

Referring to the example of the dice scores again: $\mathrm{Var}(X) = \tfrac{91}{6} - (\tfrac{21}{6})^2 = \tfrac{35}{12}$.

The expectation of a linear function of X can be expressed in terms of E(X).

In general: $E(aX + b) = aE(X) + b.$

Once the value of $E(X)$ is known, applying the result above is much simpler than working out $\Sigma\,(ax + b)p(x)$. However, the result only holds for linear functions, so for example $E(X^2)$ is not the same as $(E(X))^2$.

The variance of a linear function of X may be expressed in terms of Var(X).

In general: $\mathrm{Var}(aX + b) = a^2\,\mathrm{Var}(X).$

The binomial probability distribution

AQA	S1
EDEXCEL	S2
OCR	S1
WJEC	S1
NICCEA	S1

Suppose that n independent trials of an experiment are carried out, each with a fixed probability of success p and corresponding probability of failure q. Then $p + q = 1$ and the probability of r successes is given by the $(r + 1)$th term in the binomial expansion of $(p + q)^n$.

(You need to check that these conditions are satisfied before applying any of the results.)

Using the random variable X to represent the number of successes in n trials, the probability function is given by $P(X = r) = \binom{n}{r} p^r q^{n-r}$ $r = 0, 1, 2, ..., n.$

With this notation, the binomial probability distribution may be written as

r:	0	1	2	...	n
$P(X = r)$:	q^n	npq^{n-1}	$\dfrac{n(n-1)}{2}p^2 q^{n-2}$	...	$p^n.$

If X has a binomial distribution with parameters n and p then this is written as $X \sim B(n, p)$.

In general, if $X \sim B(n, p)$ then $\mu = E(X) = np$ and $\mathrm{Var}(X) = np(1 - p)$.

Example
The random variable $X \sim B(8, 0.4)$ find:

(a) $P(X = 2)$ (b) $P(X \leqslant 2)$ (c) $E(X)$ (d) $\mathrm{Var}(X)$.

(a) $P(X = 2) = \binom{8}{2}(0.4)^2(0.6)^6 = 0.2090 ... = 0.209$ to 3 d.p.

(b) $P(X \leqslant 2) = P(X = 0) + P(X = 1) + P(X = 2)$

$= (0.6)^{10} + 10(0.4)(0.6)^9 + 0.2090... = 0.25537...$

$= 0.255$ to 3 d.p.

$P(X \leqslant r)$ may also be found using cumulative Binomial distribution tables. This is particularly useful when r is large.

(c) $E(X) = np = 8 \times 0.4 = 3.2.$

(d) $Var(X) = np(1 - p) = 8 \times 0.4 \times 0.6 = 1.92.$

The Poisson distribution

AQA	S1
EDEXCEL	S2
OCR	S2
WJEC	S1
NICCEA	S1

The **Poisson distribution** is used to model the number of occurrences of an event in some fixed interval of space or time. It has a single parameter λ which represents the mean number of occurrences in an interval of a particular size. The events occur independently of each other and at random.

If X represents the number of occurrences of the event in an interval of a particular size then:

$$P(X = r) = \frac{e^{-\lambda}\lambda^r}{r!} \qquad r = 0, 1, 2, 3, ...$$
$$\mu = E(X) = \lambda$$
$$\sigma^2 = Var(X) = \lambda.$$

If X has a Poisson distribution with parameter λ then this is written as $X \sim Po(\lambda)$.

Example

The mean number of letters received by a household each day from Monday to Saturday is 3. Find the probability that, on a particular weekday, the number of letters received is (a) 0 (b) at least 2.

If X is the number of letters received in a day then $X \sim Po(3)$.

(a) $P(X = 0) = e^{-3} = 0.0497... = 0.050$ to 3 d.p.

(b) $P(X \geqslant 2) = 1 - (P(X = 0) + P(X = 1)) = 1 - (e^{-3} + 3e^{-3}) = 1 - 4e^{-3} = 0.8008 ...$
$= 0.801$ to 3 d.p.

The geometric distribution

| OCR | S1 |

The **geometric distribution** is used to model the number of independent trials needed before a particular outcome occurs. Taking X to represent this value and p to be the fixed probability that the outcome occurs:

$$P(X = x) = p(1 - p)^{x-1} \qquad \text{for} \quad x = 1, 2, 3, ...$$

Progress check

1 The random variable X has the probability distribution shown. x: 0   1   2
Find the value of k. $p(x)$: 0.3 0.2 k

2 Find (a) $E(X)$ (b) $Var(X)$ for the probability distribution in question 1.

3 The random variable Y is given by $Y = 3X + 2$, where X is as in question 1.
Find (a) $E(Y)$ (b) $Var(Y)$.

4 Given that $X \sim B(20, 0.3)$ write down (a) $E(X)$ (b) $Var(X)$.

4 (a) 6   (b) 4.2
3 (a) 5.6   (b) 6.84
2 (a) 1.2   (b) 0.76
1 $\lambda = 0.5$

4.4 The Normal distribution

After studying this section you should be able to:

- *understand the concept of continuous random variables*
- *use the Normal distribution model to calculate probabilities*

LEARNING SUMMARY

Continuous random variables

AQA	S1
EDEXCEL	S1
OCR	S2
WJEC	S2
NICCEA	S1

A **continuous random variable** X may take any value in some specified interval. A special function called a **probability density function** (p.d.f.) is needed to describe how the probability is distributed over this interval.

A p.d.f. is usually denoted by $f(x)$ where $f(x) \geq 0$ for all values of x.

Two important properties of a p.d.f are: $\int_{-\infty}^{\infty} f(x)\, dx = 1$

and $\quad P(a \leq X \leq b) = \int_{a}^{b} f(x)\, dx.$

> The diagram shows that values towards the upper end of the interval occur with greater likelihood than values at the lower end.

Example

A random variable has p.d.f. $f(x)$ given by:

$$f(x) = \begin{cases} kx & 1 \leq x \leq 3 \\ 0 & \text{otherwise.} \end{cases}$$

Find the value of k.

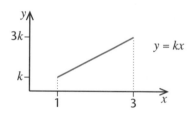

> Since $f(x)$ is linear there is no need to use integration to find the area.

The area under the graph is given by $\dfrac{3-1}{2}(k + 3k) = 4k.$

Since $f(x)$ is a p.d.f. $4k = 1 \Rightarrow k = 0.25.$

The Normal distribution

AQA	S1
EDEXCEL	S1
OCR	S2
WJEC	S2
NICCEA	S1

The p.d.f. for the **Normal distribution** is represented by a bell-shaped curve, symmetric about the mean value μ.

The parameters of the normal distribution are μ and σ^2. If the continuous random variable X follows a Normal distribution with these parameters then this is written as $X \sim N(\mu, \sigma^2)$.

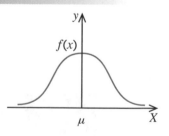

> The formula for the p.d.f. is complicated – you don't need to know it.

The **standard normal variable** Z is used to calculate probabilities based on the

> Tabulated values of probabilities for the Normal distribution use Z as the continuous random variable.

Normal distribution where $Z = \dfrac{X - \mu}{\sigma}$ and $Z \sim N(0, 1)$.

Example

Find $P(15 < X < 17)$ where $X \sim N(15, 25)$

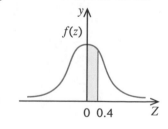

$$15 < X < 17 \Rightarrow \frac{15-15}{5} < Z < \frac{17-15}{5} \Rightarrow 0 < Z < 0.4.$$

$$P(15 < X < 17) = P(0 < Z < 0.4)$$
$$= \Phi(0.4) - \Phi(0) = 0.6554 - 0.5000$$
$$= 0.1554.$$

You may need to use given information to set up and solve simultaneous equations to find the values of μ and σ.

Example
Find P($X > 70$) = 0.0436 and   P($X < 45$) = 0.3669 where ($X \sim N(\mu, \sigma^2)$). Find the values of μ and σ.

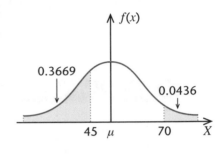

$$P(X > 70) = P\left(Z > \frac{70-\mu}{\sigma}\right) = 0.0436$$

$$\Phi\left(\frac{70-\mu}{\rho}\right) = 1 - 0.0436 = 0.9564$$

From the tables

$$\left(\frac{70-\mu}{\sigma}\right) = 1.71 \Rightarrow 70 - \mu = 1{,}71\sigma \qquad [1]$$

Similarly

$$P(X < 45) = P\left(Z < \frac{45-\mu}{\sigma}\right) = 0.3669.$$

$$\Phi\left(\frac{45-\mu}{\sigma}\right) = 0.3669$$

In general,
$\Phi(-z) = 1 - \Phi(z)$.

$$\Phi\left(-\left(\frac{45-\mu}{\sigma}\right)\right) = 1 - 0.3669 = 0.6331$$

$$-\left(\frac{45-\mu}{\sigma}\right) = 0.34$$

$$45 - \mu = -0.34\sigma \qquad [2]$$

[1] − [2] gives $25 = 2.05\sigma \Rightarrow \sigma = 12.2$

$$\mu = 49.1$$

Progress check

Find P($17 < X < 24$) where $X \sim N(14, 16)$.

0.2204

4.5 Correlation and regression

After studying this section you should be able to:

- calculate and interpret the product moment correlation coefficient
- calculate and interpret Spearman's coefficient of rank correlation
- recognise explanatory and response variables
- find the equations of least squares regression lines
- use a suitable regression line to estimate a value

LEARNING SUMMARY

All of the work in this section relates to the treatment of **bivariate data**. This is data in which each data point is defined by two variables.

Correlation

AQA	S1
EDEXCEL	S1
OCR	S1
WJEC	S3
NICCEA	S2

You should only use this method when both variables are normally distributed.

A **scatter diagram** may be used to represent bivariate data. The extent to which the points approximate to a straight line gives an indication of the strength of a linear relationship between the variables, known as the **linear correlation**.

One way to arrive at a numerical measure of the correlation is to use the **product–moment correlation coefficient**, r.

For n pairs of (x, y) values:

$$S_{xx} = \sum x^2 - n\bar{x}^2 \qquad S_{yy} = \sum y^2 - n\bar{y}^2 \qquad S_{xy} = \sum xy - n\bar{x}\bar{y},$$

You may be able to obtain the value of r directly from your calculator.

and the product–moment correlation coefficient is given by:

$$r = \frac{S_{xy}}{\sqrt{S_{xx}S_{yy}}}.$$

This gives values of r between -1 (representing a perfect negative correlation) and $+1$ (representing a perfect positive correlation).

This method may be used even when the variables are not normally distributed.

An alternative measure of correlation is given by **Spearman's rank correlation coefficient**, r_s.

The set of values for each variable must first be ranked from largest to smallest. Some care is needed with equal values:

The three equal values occupy positions 2, 3 and 4. The average positional value is given by:

$$\frac{2 + 3 + 4}{3} = 3.$$

x	rank
39	1
37	3
37	3
37	3
32	5

Each of the equal values is given a rank of 3.

At each data point, the difference in the rank of the two variables is denoted by d.

The rank correlation coefficient is given by $r_s = 1 - \dfrac{6\sum d^2}{n^3 - n}$.

This gives values of r_s between -1 (representing a perfect negative correlation) and $+1$ (representing a perfect positive correlation).

The two correlation coefficients given above are comparable but not necessarily equal.

Regression

AQA   S1
EDEXCEL   S1
OCR   S1
NICCEA   S2

Whereas correlation is determined by the *strength* of a linear relationship between the two variables, **regression** is about the *form* of the relationship given by the equation of a **regression line**. (Make sure you known the difference between correlation and regression.)

The purpose in establishing the equation of a regression line is to make predictions about the values of one variable (known as the **response variable**) for some given values of the other variable (known as the **explanatory variable**).

Predictions should only be made for values within the range of readings of the explanatory variable. **Extrapolation** for values outside this range is unreliable. Another factor affecting the accuracy of any predictions is the influence of outliers on the equation of the regression line.

Figure 1 illustrates *y*-**residuals** given by

$$d = (\text{observed value of } y) - (\text{predicted value of } y).$$

Figure 2 illustrates *x*-**residuals** given by

$$d = (\text{observed value of } x) - (\text{predicted value of } x).$$

A **least squares regression line** is a line for which the sum of the squares of either the *x*-residuals or *y*-residuals is minimised.

Figure 1

Figure 2

> Both lines always pass through the point $(\bar{x}, \bar{y})$.

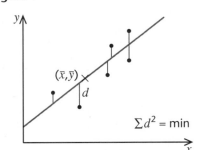

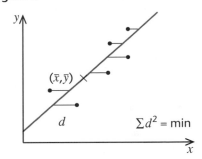

This gives the regression line of *y* on *x* as

$$y = a + bx,$$

where $b = \dfrac{S_{xy}}{S_{xx}}$ and $a = \bar{y} - b\bar{x}$.

> The values of *a* and *b* may be obtained directly from some calculators.

Use this equation to estimate values of *y* for given values of *x*. In this case, *x* is the explanatory variable and *y* is the response variable.

This gives the regression line of *x* on *y* as

$$x = c + dy,$$

where $d = \dfrac{S_{xy}}{S_{yy}}$ and $c = \bar{x} - d\bar{y}$.

Use this equation to estimate values of *x* for given values of *y*. In this case, *y* is the explanatory variable and *x* is the response variable.

Unless there is perfect correlation between the variables, the two regression lines will be different and you cannot rearrange one equation to obtain the other.

Progress check

The summary data for 10 pairs of (x, y) values is as follows:

$$\sum x = 146, \quad \sum x^2 = 2208, \quad \sum y = 147, \quad \sum y^2 = 2247, \quad \sum xy = 2211.$$

1 Find the value of the product–moment correlation coefficient between x and y.
2 Find the equation of the least squares regression line of y on x.

2 $y = 2.317 + 0.8481x$
1 0.799

Sample questions and model answers

1

Paul swims 20 lengths of a swimming pool in training for a competition. His times, x seconds, for completing lengths of the pool are summarised by:

$$\Sigma x = 372 \qquad \Sigma x^2 = 6952.$$

(a) Find the mean and standard deviation of Paul's times for completing one length of the pool.

(b) A month ago, Paul's mean time per length was 19.8 s with a standard deviation of 1.72 s. Comment on the change in his performance indicated by these results.

(a) $\qquad \bar{x} = \dfrac{\Sigma x}{n} = \dfrac{372}{20}$

$\qquad\qquad = 18.6.$

Paul's mean time per length is 18.6 s.

$$\sqrt{\frac{\Sigma x^2}{n} - \bar{x}^2} = \sqrt{\frac{6952}{20} - 18.6^2}$$

$$= 1.28.$$

The standard deviation of Paul's times is 1.28 s.

(b) The mean time has been reduced, suggesting an improvement in performance. The standard deviation has also been reduced which suggests a greater consistency of performance.

2

The letters of the word ASSOCIATES are jumbled up and arranged in a straight line.

(a) How many different arrangements are there?

(b) How many of the arrangements start with O and end with I?

(c) Find the number of arrangements in which the vowels are grouped together and the probability that this occurs.

(a) There are two As, three Ss and one of each of the other letters. There are 10 letters altogether.

The number of arrangements is given by

$$\frac{10!}{2! \; 3!} = 302\,400$$

Sample questions and model answers *(continued)*

Filling positions is a useful way to think about the problem.

(b) Fixing one letter at each end leaves eight positions to fill. This includes two As and three Ss as before.

The number of arrangements is given by

$$\frac{8!}{2! \; 3!} = 3360.$$

(c) There are six vowels, AAAEIO, that will move together and effectively occupy one position.

You need to think about the arrangements of vowels within their group.

These may be arranged in $\frac{6!}{3!} = 120$ ways.

The vowels together with three Ss and one T occupy five positions.

You also need to think of the vowels as a single unit or 'letter' to be arranged with the remaining letters.

These may be arranged in $\frac{5!}{3!} = 20$ ways.

The total number of arrangements that group the vowels together is $120 \times 20 = 2400$.

The probability that this occurs is $\dfrac{2400}{302\;400} = \dfrac{1}{126}$.

To find the total number of arrangements you need to multiply the results together.

3

A machine produces nails with lengths that are normally distributed with mean 5.2 cm and standard deviation 0.06 cm.

Find the probability that a nail selected at random has a length between 5.1 cm and 5.25 cm.

Define the random variable X.

Let X represent the continuous random variable given by the length of the nails.

Express the probability in terms of X and then standardise.

$$P(5.1 < X < 5.25) = P\left(\frac{5.1 - 5.2}{0.06} < Z < \frac{5.25 - 5.2}{0.06}\right)$$

$$= P(-1.667 < Z < 0.8333)$$

$$= \Phi(0.8333) - \Phi(-1.667)$$

$$= \Phi(0.8333) - (1 - \Phi(1.667))$$

It follows from the symmetry of the distribution that

$$\Phi(-z) = 1 - \Phi(z).$$

$$= \Phi(0.8333) + \Phi(1.667) - 1$$

$$= 0.7977 + 0.9522 - 1$$

$$= 0.7499.$$

The probability that a nail selected at random has a length between 5.1 cm and 5.25 cm is 0.7499.

Practice examination questions

1 The letters in the word COPPER are written on separate pieces of card placed on a table.
A letter is chosen by picking up the card on which it is written.

   (a) Find the number of different selections of four letters that can be made.

   (b) Find the number of different arrangements of four letters that can be made.

2 A discrete random variable X has the probability function p(x) given by

$$p(x) = \frac{x^2}{k} \quad \text{for} \quad x = 0, 1, 2, 3.$$

$$p(x) = 0 \quad \text{otherwise.}$$

   (a) Find the value of k.

   (b) Calculate the mean and variance of the probability distribution.

   (c) Find the mean and variance of the random variable $Y = 3X - 5$.

3 X is a random variable such that $X \sim B(20, 0.4)$.

   (a) Find the mean and variance of X.

   (b) Find the probability that the value of X lies between 2 and 4 inclusive.

   (c) Find the probability that X is greater than or equal to 5.

4 The events A and B are such that

$$P(A) = 0.3 \qquad P(B) = 0.6 \qquad P(A \cup B) = 0.72.$$

   (a) Show that the events A and B are not mutually exclusive.

   (b) Show that the events A and B are independent.

   (c) Write down P($A \cup B$)'.

   (d) Find P($A' \cup B'$).

Practice examination questions (continued)

5 (a) Three cards are drawn at random, without replacement, from a pack of playing cards.
Calculate the probability that:

 (i) All three cards are red.

 (ii) All three cards are the same colour.

 (iii) At least one of the cards is red.

 (b) The cards are returned to the pack and the pack is shuffled. One more card is drawn.
Find the probability that this card is an ace, given that it is not a diamond.

6 A transport company has to deal with X breakdowns of its vehicles per week.
X has a Poisson distribution with mean 0.8.

 (a) Find the probability that in one particular week:

 (i) There will be no breakdowns.

 (ii) There will be at most two breakdowns.

 (b) Find the probability that, in a four-week period, there will be no breakdowns.

7 The lifetime of batteries used in a torch can be modelled by a Normal distribution with a mean of 8 h and a standard deviation of 25 mins.

 (a) Calculate the probability that a battery selected at random will last for more than $8\frac{1}{2}$ h.

 (b) Find the percentage of batteries that will last for less than 7 h.

8 Two teachers independently mark a sample of five pieces of students work as part of a standardisation procedure. The table shows how the marks were allocated.

Sample No	1	2	3	4	5
Teacher A	39	72	51	49	68
Teacher B	46	80	53	54	73

 (a) Calculate Spearman's rank correlation coefficient for the two sets of marks.

 (b) Does a high positive correlation between two such sets of marks mean that the scripts have necessarily been marked to the same standard?
Explain your answer.

9 A machine distributes 'lucky pendants' to one in four packets of cereal at random. A box contains 20 packets of the cereal.

(a) Using tables of cumulative binomial probabilities, or otherwise, find the probability that in a randomly selected box:

   (i) At least six packets contain a lucky pendant.
   (ii) Exactly six packets contain a lucky pendant.

(a) Five boxes are randomly selected. Find the probability that exactly two of the boxes have at least six packets that contain a lucky pendant.

10 Twelve students in the sixth form of a school study both mathematics and physics at A Level.
The scores in an end of term test are shown for 10 of these students.

Maths	48	39	52	55	71	62	76	54	61	58
Physics	41	35	50	47	63	48	70	45	46	39

Using x to represent the score in maths and y to represent the corresponding score in physics:

$$x = 576 \quad y = 484 \quad x^2 = 34\,216 \quad y^2 = 24\,450 \quad xy = 28785$$

(a) Calculate the product–moment correlation coefficient, r, for the data.

(b) Find the equation of the regression line of maths scores on physics scores.

(c) Two students took the physics test but missed the maths test.

   (i) Estimate the maths score for one of these students who obtained a physics score of 42.

   (ii) Comment on the validity of estimating the maths score of the second student who obtained a physics score of 80.

Chapter 5

Decision Mathematics 1

The following topics are covered in this chapter:

- Algorithms
- Graphs and networks
- Critical path analysis

- Linear programming
- Matchings

5.1 Algorithms

After studying this section you should be able to:

- *implement an algorithm given by text or flow chart*
- *implement the following algorithms from memory:*
  - *bubble sort*
  - *quick sort*
  - *binary search*
  - *bin packing*

LEARNING SUMMARY

Introducing algorithms

AQA	D1
EDEXCEL	D1
OCR	D1

Algorithms play a fundamental role throughout Decision Mathematics.

In the exam you may be asked to implement an unfamiliar algorithm and interpret the results.

An **algorithm** is a finite sequence of precise instructions used to solve a problem.

One way to define an algorithm is to simply list all of the necessary steps. In more complex situations, the process may be easier to follow if the algorithm is presented as a **flow diagram**.

You need to be able to implement an algorithm presented in either form.

Example
Implement the algorithm given by the flow chart and state what the written values represent.

The table shows how the values of I and J change as the algorithm is implemented.

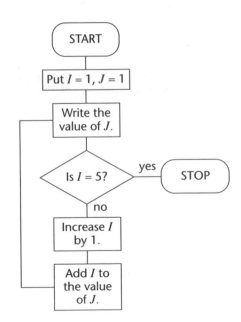

I	1	2	3	4	5
J	1	3	6	10	15

The written values are 1, 3, 6, 10, 15 and these are the first five triangle numbers.

The algorithm may be expressed in words as:

Write down a sequence of 5 numbers starting with 1 and then increasing, first by 2 and then by 1 more each time from term to term.

This version is more concise but not so easy to understand.

In the exam you may be expected to implement any of these algorithms without being reminded of the necessary steps.

You need to know and be able to implement the following algorithms.

The bubble sort

EDEXCEL D1
OCR D1

As its name suggests, the **bubble sort** is an algorithm for sorting a list in a particular order.

Step 1 Compare the first two elements of the list and switch them if they are in the wrong order.

Step 2 Compare the second and third elements of the list and, again, switch them if they are in the wrong order.

Step 3 Continue in the same way until you reach the end of the list. This completes the first **pass** through the list.

Step 4 Make repeated passes through the list until a pass produces no change.

Example   Sort the list 5, 7, 3, 11, 6, 8 into ascending order.

The first pass gives

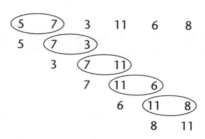

The second pass produces   3, 5, 6, 7, 8, 11.
The third pass produces   3, 5, 6, 7, 8, 11.

The third pass made no change so the ordering is complete.

Don't just look at the list and say that the numbers are now in order. Use the algorithm to decide when to stop.

The quick sort

EDEXCEL D1
OCR D1

Step 1 Take the first element in the list as the **pivot**.

Step 2 Consider each of the other elements in turn and place any with order less than or equal to the pivot on its one side and any with order greater than the pivot on its other side. Do this without re-ordering the numbers on either side of the pivot.

Step 3 Repeat steps 1 and 2 for each sub-list until each one contains a single element.

This produces two sub-lists with the pivot in-between.

Using the quick sort algorithm on the same list as the last example gives:

$$3 \quad \underline{5} \quad 7 \quad 11 \quad 6 \quad 8$$
$$6 \quad 7 \quad 11 \quad 8$$
$$8 \quad 11$$

Build the ordered list using the values to the left of each pivot and the pivot itself.

The ordered list is 3, 5, 6, 7, 8, 11.

Binary search

AQA	D1
EDEXCEL	D1
OCR	D1

The binary search algorithm provides a systematic way of searching through an ordered list to find an element that matches your criterion. **For example,** you may need to look through a set of index cards to find the contact details of a customer.

 This is not a very efficient method if the required element is at the end of the list.

Step 1   Find the element in the middle of the list. If this element matches your criterion then stop. If it does not, then use the middle term to decide in which half of the list the required element must lie.

Step 2   Repeat Step 1 for the selected half.

Bin packing

AQA	D1
EDEXCEL	D1
OCR	D1

The bin packing algorithm is used to solve problems that can be represented by the need to pack some boxes of equal cross-section but different heights into bins, with the same cross-section as the boxes, using as few bins as possible.

There is no known algorithm that will always provide the *best* solution. You need to be familiar with the three algorithms below that attempt to provide a *good* solution. Such algorithms are known as **heuristic** algorithms.

First-fit algorithm

Take each box in turn from the order given and pack it into the first available bin.

Full-bin algorithm

The full-bin algorithm is only practical when the number of bins and boxes is small.

Step 1   Considering the boxes in the given order, use the first available combination that will fill a bin.

Step 2   Repeat Step 1 until no more bins can be filled.

Step 3   Implement the first-fit algorithm for the remaining boxes.

First-fit decreasing algorithm

Step 1   Arrange the boxes in decreasing order of size.

Step 2   Implement the first-fit algorithm starting with the largest box.

Progress check

1   Write the letters P, Q, B, C, A, D in alphabetical order using:

(a) the bubble sort algorithm

(b) the quick sort algorithm.

2   A project involves activities A–H with durations in hours as given in the table.

A	B	C	D	E	F	G	H
3	1	5	4	2	3	4	2

You can represent each worker as a 'bin' of 'height' 8 hours.

The project is to be completed in 8 hours.

(a) Use the first-fit algorithm to try to find the minimum number of workers needed.

(b) Find a better solution.

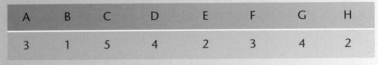
2   First-fit: 4 workers; A better solution is (A, B, D) (C, F) (E, G, H).

1   A, B, C, D, P, Q.

5.2 Graphs and networks

After studying this section you should be able to:

- *understand the use of terminology associated with graphs and networks*
- *use Prim's and Kruskal's algorithms for constructing a minimum spanning tree*
- *use Dijkstra's algorithm to find a shortest path between two vertices*
- *find upper and lower bounds for the travelling salesman problem*
- *use the route inspection algorithm to find a route of minimum weight that traverses every arc of a network*
- *find the maximum flow in a network with directed edges*

LEARNING SUMMARY

Defining terms

> It is worth spending time getting to know all of the terms. Their definitions are quite detailed so you will need to keep reminding yourself of what each term means.

- A **graph** is a set of points, called **vertices** or **nodes**, connected by lines called **edges** or **arcs**.
- A **simple graph** is one that has no loops and in which no pair of vertices are connected by more than one edge.

A simple graph   A non-simple graph   A non-simple graph

Two edges between A and B   A loop around C

- The number of edges incident on a vertex is called its **order**, **degree** or **valency**. A vertex may be **odd** or **even** depending on whether its order is odd or even.
- A **subgraph** of some graph G is a graph consisting entirely of vertices and edges that belong to G.
- A **directed** edge is an edge that has an associated direction shown by an arrow. A **digraph** is a graph in which the edges are directed.
- A **path** is a finite sequence of edges such that the end vertex of one edge is the start vertex of the next edge. No edge is included more than once.
- A **connected graph** is one in which every pair of vertices is connected by a path.

  A **complete graph** is one in which every pair of vertices is connected by an edge. A complete graph with n vertices is denoted by K_n.
- A **planar graph** is one that can be drawn in a plane such that no two edges meet except at a vertex.

  K_4 is planar but K_5 is not planar.

K_4 K_5

- A **cycle** is a path that starts and finishes at the same vertex.
- An **Eulerian cycle** is one that traverses all of the edges of the graph.
- A **Hamiltonian cycle** is one that passes through every vertex once only.
- A **tree** is a graph with no cycles.
- A **spanning tree** is a tree whose vertices are all of the vertices of the graph.
- A **network** is a graph that has a number (**weight**) associated with every edge.
- A **minimum spanning tree (MST)** of a network is a spanning tree of minimum possible weight. It is sometimes called a **minimum connector**.

Prim's algorithm

AQA   D1
EDEXCEL   D1
OCR   D1

Prim's algorithm may be used to find a minimum spanning tree of a network.

Step 1   Choose a starting vertex.

Step 2   Connect it to the vertex that will make an edge of minimum weight.

Step 3   Connect one of the remaining vertices to the tree formed so far, in such a way that the minimum extra weight is added to the tree.

Step 4   Repeat step 3 until all of the vertices are connected.

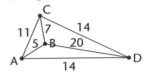

Starting from A gives the MST as

> The number in row A and column B, for example, represents the weight of the edge AB.

The network given above may be represented by the matrix shown.

A version of Prim's algorithm may be used to find the minimum spanning tree directly from a matrix.

Step 1   Choose a starting vertex. Delete the corresponding row and write 1 above the corresponding column as a label.

Step 2   Circle the smallest undeleted value in the labelled column and delete the row in which it lies.

Step 3   Label the column, corresponding to the vertex of the deleted row, with the next label number.

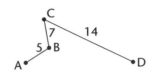

AD could have been used instead of CD

	A	B	C	D
A	–	5	–	–
B	5	–	7	–
C	–	7	–	14
D	–	–	14	–

> If there is more than one smallest value then you can choose which one to circle.

Step 4   Circle the smallest undeleted value of all the values in the labelled columns and delete the row in which it lies.

Step 5   Repeat steps 3 and 4 until all of the rows are deleted. The circled values then define the edges of the minimum spanning tree.

The circled values correspond to the edges AB, BC, and CD.

Kruskal's algorithm

AQA   D1
EDEXCEL   D1
OCR   D1

Kruskal's algorithm is an alternative way to find a minimum spanning tree. In this case, the subgraph produced may not be connected until the final stage.

Step 1   Choose an edge with minimum weight as the first subgraph.

Step 2   Find the next edge of minimum weight that will not complete a cycle when taken with the existing subgraph. Include this edge as part of a new subgraph.

Step 3   Repeat step 2 until the subgraph makes a spanning tree.

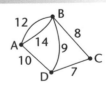

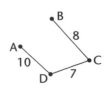

BD can not be included

Dijkstra's algorithm

AQA   D1
EDEXCEL   D1
OCR   D1

Dijkstra's algorithm is used to find the shortest distance between a chosen *start* vertex and any other vertex in a network.

The algorithm is designed for use by computers but, when implementing it without the aid of a computer, a system of labelling is required.

> The word *distance* is used here in place of the more general *weight* because in practical applications of Dijkstra's algorithm the weight so often represents a distance.

> Reading the letters inside the box as OWL might help you to remember which part is for which information.

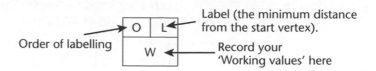

Order of labelling

O | L ← Label (the minimum distance from the start vertex).

W ← Record your 'Working values' here

The steps in the algorithm below refer to this system of labelling. Every vertex is labelled in the same way.

Step 1   Give the start vertex a label of 0 and write its order of labelling as 1.

Step 2   For each vertex directly connected to the start vertex, enter the distance from the start vertex as a working value. Enter the smallest working value as a label for that vertex. Record the order in which it was labelled as 2.

Step 3   For each vertex directly connected to the one that was last given a label, add the edge distance onto the label value to obtain a total distance. This distance becomes the working value for the vertex unless a lower one has already been found.

Step 4   Find the vertex with the smallest working value not yet labelled and label it. Record the order in which it was labelled.

Step 5   Repeat steps 3 and 4 until the target vertex is labelled. The value of the label is the minimum distance from the start vertex.

Step 6   To find the path that gives the shortest distance, start at the target vertex and work back towards the start vertex in such a way that an edge is only included if its distance equals the change in the label values.

Example
Use Dijkstra's algorithm to find the shortest distance from A to F in this network and state the route used.

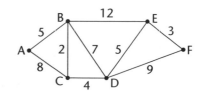

Using Dijkstra's algorithm produces the diagram on the right.

The shortest distance from A to F is 19 (given by the label at F).

Tracing the route backwards using Step 6 gives F, E, D, C, B, A, so the required route is A, B, C, D, E, F.

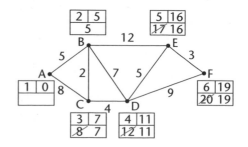

The travelling salesman problem

AQA   D2
EDEXCEL   D1
OCR   D1

This is based on the classic situation of a salesman who wishes to visit a number of towns and return home using the shortest possible route.

The **travelling salesman problem** (TSP) is the problem of finding a route of minimum distance that visits every vertex and returns to the start vertex. For a small network it is possible to produce an exhaustive list of all possible routes and choose the one that minimises the total distance. For a large network an exhaustive check is not feasible, even with a computer, because the number of possible routes grows so rapidly.

It is useful to know within what limits the total distance must lie and there are algorithms that can be used for this purpose.

Finding an upper bound

> **KEY POINT**
>
> An upper bound for the total distance involved in the travelling salesman problem is given by twice the minimum spanning tree.

The minimum spanning tree (MST) for the network ABCD was found earlier.

The total length of the MST is $5 + 7 + 14 = 26$.

An upper bound for the TSP is $2 \times 26 = 52$. This corresponds to the route ABCDCBA.

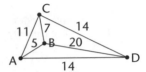

In some cases you might be able to find more than one short-cut that will enable you to reduce the upper bound.

However, an improved (smaller) upper bound can be found by using a **short-cut** from D directly back to A. The route is then ABCDA and the corresponding upper bound is 40.

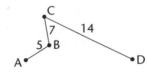

A different approach is to use the **nearest neighbour algorithm**.

Step 1   Choose a starting vertex.

Step 2   Move from your present position to the nearest vertex not yet visited.

Step 3   Repeat step 2 until every vertex has been visited. Return to the start vertex.

For the network ABCD considered above, the nearest neighbour algorithm gives the same result as using the MST with a short-cut. This won't always be the case.

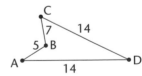

Finding a lower bound

Step 1   Choose a vertex and delete it along with all of the edges connected to it.

Step 2   Find a minimum spanning tree for the remaining part of the network.

Step 3   Add the length of the MST to the lengths of the two shortest deleted edges.

The value found at step 3 is a lower bound for the travelling salesman problem. It may be possible to find a better (larger) lower bound by choosing to delete a different vertex initially and repeating the process.

Deleting D to start with gives a lower bound of 40. This corresponds to a cycle ABCDA and so represents the solution of the TSP.

Using the network ABCD again and deleting A gives the minimum spanning tree BCD of length $7 + 14 = 21$. Adding the two shortest deleted lengths gives a lower bound of 37.

The route inspection problem

AQA   D1
EDEXCEL   D2
OCR   D2

Again, it is common to use length rather than weight in this context.

The **route inspection problem** is to find a route of minimum total length that traverses every edge of the network, at least once, and returns to the start vertex.

This is also known as the Chinese postman problem.

The algorithm for solving this problem is based on the idea of a **traversable** graph. A graph is traversable if it can be drawn in one continuous movement without going over the same edge more than once. If all of the vertices of the graph are even, then any one of them may be used as the starting point and the same point will be returned to at the end of the movement. Such a graph is said to be **Eulerian**. The only other possibility for a traversable graph is that it has exactly two odd vertices. In this situation, one of the vertices is the start point and the other is the finish point. This type of graph is said to be **Semi-Eulerian**.

Leonhard Euler was the most prolific mathematician of all time. One of his many contributions to the subject was to introduce graph theory as a means of solving problems.

The **route inspection algorithm** is:

Step 1   List all of the odd vertices.

Step 2   Form the list into a set of pairs of odd vertices. Find all such sets.

Step 3   Choose a set. For each pair find a path of minimum length that joins them. Find the total length of these paths for the chosen set.

Step 4   Repeat steps 3 until all sets have been considered.

Repeating the edges between the pairs of odd vertices in this way effectively makes all the vertices even and creates a traversable graph. See page 126 for an example.

Step 5   Choose the set that gives the minimum total. Each pair in the set defines an edge that must be repeated in order to solve the problem.

Flows in networks

AQA   D1
EDEXCEL   D1
OCR   D2

The **flows** referred to may be flows of liquids, gases or any other measurable quantities. The edges may represent such things as pipes, wires or roads that carry the quantities between the points identified as vertices.

A typical vertex has a flow into it and a flow out of it. The exceptions are a **source** vertex which has no input and a **sink** vertex which has no output.

Each edge of the network has a **capacity** which represents the maximum possible flow along that edge. In the usual notation, the capacity is written next to the edge and the flow is shown in a circle.

This edge has a capacity of 10 and a flow of 7.

The set of flows for a network is **feasible** if:

• The total output from all source vertices is equal to the total input for all sink vertices (most will have just one source vertex and one sink vertex).

• The input for each vertex other than a source or sink vertex is equal to its output.

• The flow along each edge is less than or equal to its capacity.

A **cut** divides the vertices into two sets, one set containing the source and the other containing the sink.

The **capacity of a cut** is equal to the sum of the capacities of the edges that cross the cut, *taken in the direction from the source set to the sink set*.

In the diagram, the capacity of cut (i) is $15 + 10 = 25$.

The capacity of cut (ii) is $8 + 14 = 22$.

Be careful to add the capacities and not the flows.

Notice that, for the second cut, the capacity of 4 is not included because it does not go from the source set to the sink set.

The maximum flow–minimum cut theorem

AQA	D1
EDEXCEL	D1
OCR	D1

The **maximum flow–minimum cut theorem** states that the maximum value of the flow through a network is equal to the capacity of the minimum cut. (This is a little bit like saying that a chain is only as strong as its weakest link.)

It follows from the theorem that if you find a flow that is equal to the capacity of some cut then it must be the maximum flow that can be established through the network.

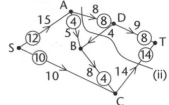

For example, the capacity of cut (ii) for this network has already been found to be 22. The circled figures show one way of achieving this flow and the theorem now tells us that this must be the maximum possible flow through the network.

Flow augmentation

AQA	D1
EDEXCEL	D1
OCR	D2

Once an initial flow has been established through a network it is useful to know the extent to which the flow along any edge may be altered in either direction.

This is labelled as **excess capacity** and **back capacity** on each edge as shown. It may be possible to **augment** the flow through the network by making use of this flexibility along a path from the source vertex to the sink vertex.
Such a path is called a **flow-augmenting path**.

An algorithm for finding the maximum flow through a network by augmentation is:

Step 1   Find an initial flow by inspection.

Step 2   Label the excess capacity and back capacity for each edge.

Step 3   Search for a flow-augmenting path. If one can be found then increase the flow along the path by the maximum amount that remains feasible.

Step 4   Repeat steps 2 and 3 until no flow-augmenting path may be found.

Progress check

1. (a) Use Prim's algorithm to find the total weight of a minimum spanning tree for this network.

    (b) Verify your answer to part (a) using Kruskal's algorithm.

2. Use Dijkstra's algorithm to find the shortest distance from A to G in the network given in question 1. State the route used.

3. Find the maximum flow through this network. Verify your answer using the maximum flow–minimum cut theorem.

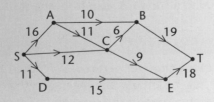

5.3 Critical path analysis

After studying this section you should be able to:

- *construct an activity network from a given precedence table including the use of dummies where necessary*
- *use forward and backward scans to determine earliest and latest event times*
- *find the critical path in an activity network*
- *calculate the total float of an activity*
- *construct a chart for the purpose of scheduling*

LEARNING SUMMARY

The process of representing a complex project by a network and using it to identify the most efficient way to manage its completion is called **critical path analysis**.

Activity networks

A complex project may be divided into a number of smaller parts called **activities**. The completion of one or more activities is called an **event**.

Activities often rely on the completion of others before they can be started.

The relationship between these activities can be represented in a **precedence table**, sometimes called a **dependency table**.

In the precedence table shown on the right, the figures in brackets represent the **duration** of each activity, i.e. the time required, in hours, for its completion.

Activity	Depends on
A(3)	–
B(5)	–
C(2)	A
D(3)	A
E(3)	B, D
F(5)	C, E
G(1)	C
H(2)	F, G

A precedence table can be used to produce an **activity network**. In the network, activities are represented by arcs and events are represented by vertices.

The vertices are numbered from 0 at the **start vertex** and finishing at the **terminal vertex**.

The direction of the arrows shows the order in which the activities must be completed.

There must only be *one* activity between each pair of events in the network. The notation (i, j) is used to represent the activity between events i and j.

A **dummy activity** is one that has zero duration. A dummy is needed in this network to show that G depends on C whereas F depends on C *and* E.

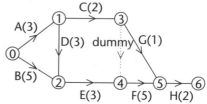

A dummy is shown with a dotted line. Its direction is important in defining dependency. In this case it shows that F depends on C not that G depends on E.

Earliest event times

The **earliest event time** for vertex i is denoted by e_i and represents the earliest time of arrival at event i with all dependent activities completed. These times are calculated using a **forward scan** from the start vertex to the terminal vertex.

Latest event times

The **latest event time** for vertex i is denoted by l_i and represents the latest time that event i may be left without extending the time for the project. These times are calculated using a **backward scan** from the terminal vertex back to the start vertex.

The **critical path** is the longest path through the network. The activities on this path are the **critical activities**. If any critical activity is delayed then this will increase the time needed to complete the project. The events on the critical path are the **critical events** and for each of these $e_i = l_i$.

It is useful to add the information about earliest and latest times to the network. The critical path is then easily identified.

Notice that B(5) lies between two critical events but it is not a critical activity.

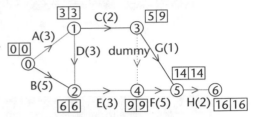

> The total float of any critical activity is always zero.

The **total float of an activity** is the maximum time that the activity may be delayed without affecting the length of the critical path. It is given by:

latest finish time − earliest start time − duration of the activity.

Scheduling

AQA D1
EDEXCEL D1
OCR D2

The process of allocating activities to workers for completion, within all of the constraints of the project, is known as **scheduling**.

The information regarding earliest and latest times for each activity is crucial when constructing a schedule. This information may be presented as a table or as a chart.

> Typically, the purpose of scheduling is to determine the number of workers needed to complete the project in a given time, or to determine the minimum time required for a given number of workers to complete the project.

Activity	Duration	Start Earliest	Start Latest	Finish Earliest	Finish Latest	Float
A(0, 1)	3	0	0	3	3	0
B(0, 2)	5	0	1	5	6	1
C(1, 3)	2	3	7	5	9	4
D(1, 2)	3	3	3	6	6	0
E(2, 4)	3	6	6	9	9	0
F(4, 5)	5	9	9	14	14	0
G(3, 5)	1	5	13	6	14	8
H(5, 6)	2	14	14	16	16	0

The critical activities are shown along one line.

The diagram illustrates the degree of flexibility in starting activities B, C and G. Remember that G cannot be started until C has been completed.

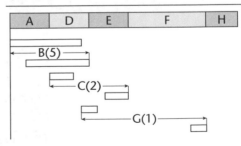

Progress check

Determine the critical activities and the length of the critical path for this network.

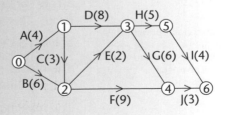

A, D, G and J, 21

121

5.4 Linear programming

After studying this section you should be able to:

- formulate a linear programming problem in terms of decision variables
- use a graphical method to represent the constraints and solve the problem
- use the Simplex algorithm to solve the problem algebraically

LEARNING SUMMARY

Formulating a linear programming problem

AQA	D1
EDEXCEL	D1
OCR	D1

x and *y* are often used for the variables.

Typically, this may be to maximise a profit or minimise a loss.

To formulate a linear programming problem you need to:

- Identify the **variables** in the problem and give each one a label.
- Express the **constraints** of the problem in terms of the variables. You need to include non-negativity constraints such as $x \geqslant 0$, $y \geqslant 0$.
- Express the quantity to be optimised in terms of the variables. The expression produced is called the **objective function**.

Example
A small company produces two types of armchair. The cost of labour and materials for the two types is shown in the table.

	Labour	Materials
Standard	£30	£25
Deluxe	£40	£50

The total spent on labour must not be more than £1150 and the total spent on materials must not be more than £1250. The profit on a standard chair is £70 and the profit on a deluxe chair is £100. How many chairs of each type should be made to maximise the profit?

In this case, the variables are the number of chairs of each type that may be produced. Using x to represent the number of standard chairs and y to represent the number of deluxe chairs, the constraints may be written as:

It's a good idea to simplify the constraints where possible.

$$30x + 40y \leqslant 1150 \Rightarrow 3x + 4y \leqslant 115$$
$$25x + 50y \leqslant 1250 \Rightarrow x + 2y \leqslant 50$$

and $\quad x \geqslant 0, \quad y \geqslant 0.$

Using P to stand for the profit, the problem is to maximise $P = 70x + 100y$.

The graphical method of solution

AQA	D1
EDEXCEL	D1
OCR	D1

Each constraint is represented by a region on the graph. It's a good idea to shade out the *unwanted* region for each one. The part that remains unshaded then defines the **feasible region** containing the points that satisfy all of the constraints.

The blue line represents the points where the profit takes a particular value. Moving the line in the direction of the arrow corresponds to increasing the profit. This suggests that the maximum profit occurs at the point X.

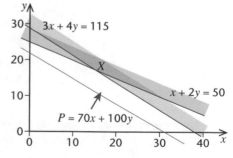

X does not represent the solution in this case because both *x* and *y* must be integers.

Solving $3x + 4y = 115$ and $x + 2y = 50$ simultaneously gives X as $(15, 17.5)$.

The nearest points with integer coordinates in the feasible region are $(15,17)$ and $(14,18)$ The profit, given by $P = 70x + 100y$, is greater at $(14,18)$.

The maximum profit is made by producing 14 standard and 18 deluxe chairs.

The Simplex algorithm

AQA   D2
EDEXCEL   D1
OCR   D1

When there are three or more variables, a different approach is needed. The **Simplex algorithm** may be used to solve the problem algebraically. The information must be expressed in the right form before the algorithm can be used.

- First write the constraints, other than the non-negativity conditions, in the form $ax + by + cz \leqslant d$.
- Write the objective function in a form which is to be maximised.
- Add **slack variables** to convert the inequalities into equations.
- Write the information in a table called the **initial tableau**.

> The objective function is already in a form which is to be maximised.

> The constraints are in the right form.

Example

Find the maximum value of $P = 2x + 3y + 4z$ subject to the constraints:

$$3x + 2y + z \leqslant 10 \qquad\qquad [\text{i}]$$

$$2x + 5y + 3z \leqslant 15 \qquad\qquad [\text{ii}]$$

$$x \geqslant 0, \quad y \geqslant 0, z \geqslant 0.$$

Using slack variables s and t, the inequalities [i] and [ii] become:

$$3x + 2y + z + s = 10$$

$$2x + 5y + 3z + t = 15.$$

The objective function must be rearranged so that all of the terms are on one side of the equation to give:

$$P - 2x - 3y - 4z = 0.$$

The information may now be put into the initial tableau.

	P	x	y	z	s	t	value
The top row shows the objective function	1	−2	−3	−4	0	0	0
This row shows the first constraint	0	3	2	1	1	0	10
This row shows the second constraint	0	2	5	3	0	1	15

The columns for P, s and t contain zeros in every row apart from one. In each case, the remaining row contains 1 and the value of the variable is given in the end column of that row. The value of every other variable is taken to be zero. So, at this stage, the tableau shows that $P = 0$, $s = 10$, $t = 15$ and x, y and z are all zero. This corresponds to the situation at the origin.

Using the algorithm is equivalent to visiting each vertex of the feasible region until an optimum solution is found. This occurs when there are no negative values in the objective row.

When the present tableau is not optimal, a new tableau is formed as follows:

- The column containing the most negative value in the objective row becomes the **pivotal column**. In this case the pivotal column corresponds to the variable z.
- Now divide each entry in the *value* column by the corresponding entry in the pivotal column provided that the pivotal column entry is positive.
  For the example above this gives $\frac{10}{1} = 10$ and $\frac{15}{3} = 5$
  The smallest of these results relates to the bottom row which is now taken to be the **pivotal row**. The entry lying in both the pivotal column and the pivotal row then becomes the **pivot**. In this case, the pivot is 3.

P	x	y	z	s	t	value	
1	−2	−3	−4	0	0	0	
0	3	2	1	1	0	10	
0	2	5	3	0	1	15	← pivotal row

pivotal column

- Divide every value in the pivotal row by the pivot, this makes the value of the pivot 1. The convention is to use fraction notation rather than decimals.

P	x	y	z	s	t	value	
1	−2	−3	−4	0	0	0	
0	3	2	1	1	0	10	
0	$\frac{2}{3}$	$\frac{5}{3}$	1	0	$\frac{1}{3}$	5	← pivotal row

- Now turn the other values in the pivotal column into zeros by adding or subtracting multiples of the pivotal row.

P	x	y	z	s	t	value	
1	$\frac{2}{3}$	$\frac{11}{3}$	0	0	$\frac{4}{3}$	20	
0	$\frac{7}{3}$	$\frac{1}{3}$	0	1	$-\frac{1}{3}$	5	
0	$\frac{2}{3}$	$\frac{5}{3}$	1	0	$\frac{1}{3}$	5	← pivotal row

> It is often necessary to repeat the process in order to find the optimum tableau.

- If there are no negative values in the objective row then the **optimum tableau** has been found. Otherwise choose a new pivotal column and repeat the process.

Each time through the process is called an **iteration**.

In this case the optimum tableau has been found after just one iteration. It shows that the maximum value of P is 20 and that this occurs when

$x = 0$, $y = 0$, $z = 5$, $s = 5$ and $t = 0$.

Progress check

An initial simplex tableau for a linear programming problem is given by:

P	x	y	z	s	t	value
1	−3	−2	−1	0	0	0
0	2	3	1	1	0	10
0	3	4	2	0	1	18

1 Carry out one iteration to produce the next tableau.

2 Explain why the tableau found is optimal. Write down the maximum value of P and the corresponding values of the other variables.

2 There are no negative values in the objective row.
$P = 15$, $x = 5$, $y = 0$, $z = 0$, $S = 0$, $t = 3$.

1	P	x	y	z	s	t	value
	1	0	$\frac{5}{2}$	$\frac{1}{2}$	$\frac{3}{2}$	0	15
	0	1	$\frac{3}{2}$	$\frac{1}{2}$	$\frac{1}{2}$	0	5
	0	0	$-\frac{1}{2}$	$\frac{1}{2}$	$-\frac{3}{2}$	1	3

5.5 Matchings

After studying this section you should be able to:

- *use bipartite graphs to model matchings*
- *understand the conditions for matchings to be maximal or complete*
- *apply the maximum matching algorithm*

Matchings and graphs

A **bipartite graph** is a graph in which the vertices are divided into two sets such that no pair of vertices in the same set is connected by an edge.

In this case, the two sets are {A, B, C} and {p, q, r, s}.

Some vertices in a bipartite graph may not be connected to another vertex.

A **matching** between two sets may be represented by a bipartite graph in which there is at most one edge connecting a pair of vertices.

A **maximal matching** is a matching which has the maximum number of edges. This occurs when every vertex in one of the sets is connected to a vertex in the other set. The bipartite graph shown above represents a maximal matching.

A **complete matching** is a matching in which every vertex is connected to another vertex. This can only occur when the two sets contain the same number of vertices.

The matching improvement algorithm

Figure 1 is a bipartite graph showing the possible connections between two sets. It does not represent a matching because some vertices have more than one connection.

Figure 2 is a bipartite graph representing an initial matching.

An initial matching may be improved by increasing the number of connections. This is the purpose of the **matching improvement algorithm**.

Figure 1   Figure 2

In an alternating path, the edges alternate between those that are not in the initial matching and those that are.

Step 1 Start from a vertex not connected in the initial matching and look for an **alternating path** to a vertex in the other set that is not connected.

Step 2 Each edge on the alternating path, not included in the initial matching, is now included and each edge originally included is removed.

Step 3 Repeat steps 1 and 2 until no further alternating paths can be found.

Alternating path   Maximal matching

Progress check

Starting with an initial matching in which 1 is connected to R and 2 is connected to P, use the improvement algorithm to establish a complete matching for this bipartite graph.

1 – P, 2 – S, 3 – R, 4 – Q.

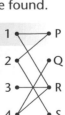

Sample questions and model answers

1

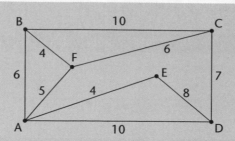

The diagram represents the system of pathways that a security guard must patrol during his course of duty. The weights on the edges represent the time taken, in minutes, to patrol each pathway.

Find the minimum time required to patrol every pathway at least once and give a possible route.

You need to recognise that this question requires the use of the route inspection algorithm.

Step 1 of the algorithm is to list the odd vertices.

Vertex	Order
A	4
B	3
C	3
D	3
E	2
F	3

The odd vertices are B, C, D and F.

The sets of pairs of odd vertices are: {BC, DF}
{BD, CF}
{BF, CD}.

Step 2 of the algorithm is to find the sets of pairs of odd vertices. Each odd vertex appears once in each set.

For the set {BC, DF}

BC = 10. The shortest route from D to F is DC + CF = 13.

There is a an edge connecting B and C.

Total length of the extra paths for this set is 10 + 13 = 23.

Step 3 of the algorithm is to find the extra length introduced for each set.

For the set {BD, CF}

The shortest route from B to D is BA + AD = 16. CF = 6.

Total length of the extra paths for this set is 16 + 6 = 22.

For the set {BF, CD}

BF = 4 and CD = 7.

There is an edge connecting C and F.

Total length of the extra paths for this set is 4 + 7 = 11.

The minimum total corresponds to the set {BF, CD} which means that the edges BF and CD will be repeated.

Step 4 of the algorithm is to keep going until all the sets have been considered.

There are many possible routes.

A possible route is ADCBAECFBFA.

The time taken will be 60 + 11 = 71 minutes.

Sample questions and model answers (continued)

2

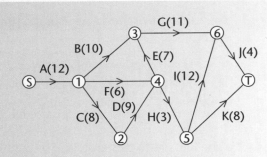

The diagram is an activity network for a project.

The figures in brackets represent the time in days to complete each activity.

The start vertex is S and the terminal vertex is T.

(a) Label each vertex with the earliest and latest time for the event. Give the length of the critical path.

(b) State which events are critical and give the critical path.

> Use a forward scan to find the earliest event times.

> Use a backward scan to find the latest event times.

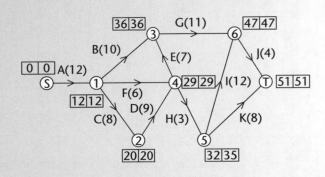

(a) The length of the critical path is 51 days.

(b) The critical events are S, 1, 2, 3, 4, 6 and T.
The critical path is A, C, D, E, G and J.

> The length of the critical path is given by the earliest (and latest) event time at the terminal vertex.

Practice examination questions

1 The table shows the amount of memory taken up by some files on a computer system.

File	A	B	C	D	E	F	G
Memory (Kb)	580	468	610	532	840	590	900

The files are to be transferred onto disks with a capacity of 1.4 Mb (1000 Kb = 1 Mb).

(a) Show that a minimum of four disks is required.

(b) Use the first-fit algorithm to allocate the programs to disks.

(c) Use the first-fit decreasing algorithm to show how all of the files may be stored using four disks.

2 A new theme park has its major attractions at A, B, C, D, E, F and G as shown in the network below. The edges of the network show the *possible* routes of paths connecting the attractions.

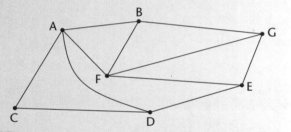

The cost of laying each of these paths, in hundreds of pounds, is given in the table:

	A	B	C	D	E	F	G
A	–	50	40	61	–	38	–
B	50	–	–	–	–	47	54
C	40	–	–	35	–	–	–
D	61	–	35	–	42	–	–
E	–	–	–	42	–	25	10
F	38	47	–	–	25	–	63
G	–	54	–	–	10	63	–

The paths that are actually laid must form a connected network.
Use Prim's algorithm, starting by deleting row A, to find the minimum cost of laying the necessary paths.

Practice examination questions (continued)

3

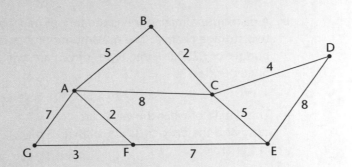

(a) Use Dijkstra's algorithm to find the shortest distance from G to D.

(b) Describe the route taken.

4

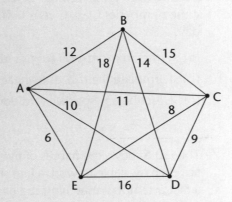

(a) Use Kruskal's algorithm to find the length of the minimum spanning tree for this network.

State the order in which you include each of the edges.

(b) Find an upper bound for the travelling salesman problem using the minimum spanning tree with shortcuts.

5

(i)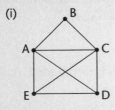

(ii)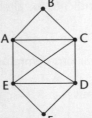

(iii)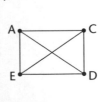

(a) List the order of the vertices for each graph.

(b) Describe each of the graphs as Eulerian, Semi-Eulerian or neither.

6 A distribution manager has the task of transporting pallets of frozen food between two storage centres. The maximum number of pallets is to be delivered within a daily budget of £2000. Three types of van are available and each van can only do the trip once per day.

The details of what the vans can carry and the daily costs are shown in the table.

At most 10 drivers may be used in one day.

Van type	No. of pallets	Cost/day
Class A	4	£120
Class B	9	£300
Class C	12	£400

Let a represent the number of Class A vans used, b the number of Class B vans used, c the number of Class C vans used and P the total number of pallets transported in a day.

(a) Write an expression for the objective function.

(b) Formulate the task as a linear programming problem.

(c) Describe any special condition that the solution must satisfy.

7 The table shows the names of five employees of a company and the days when they are each available to work a late shift. The days that have been highlighted represent a first attempt at matching the employees to the available days to make a rosta.

(a) Use a bipartite graph to represent the relationship between employees and days available.

(b) Use a second bipartite graph to represent an initial matching based on the days highlighted.

(c) Implement a matching improvement algorithm to obtain a complete matching. Indicate the alternating path used.

Name	Days available
Abbie	**Mon**, Tue, Fri
Ben	Mon, **Wed**, Thu
Colin	Mon, **Fri**
Dave	Fri, Wed
Emma	**Thu**, Fri

8 A linear programming problem is formulated as:

Maximise $\quad P = x + 2y + 3z,$

subject to $\quad 2x + 6y + z \leqslant 10$

$\qquad\qquad x + 4y + 5z \leqslant 16$

and $\quad x \geqslant 0, y \geqslant 0, z \geqslant 0.$

(a) Set up an initial simplex tableau to represent the problem.

(b) Use the simplex algorithm to solve the problem.
    State the corresponding values of x, y and z.

(c) Explain how you know that the optimum solution has been found.

Practice examination answers

Pure 1

1 (a) $49^{-\frac{1}{2}} = \dfrac{1}{49^{\frac{1}{2}}} = \dfrac{1}{\sqrt{49}} = \dfrac{1}{7}$.

(b) $\dfrac{a \times a\sqrt{a^3}}{\sqrt{a}} = \dfrac{a \times a^{\frac{3}{2}}}{a^{\frac{1}{2}}} = \dfrac{a^{\frac{5}{2}}}{a^{\frac{1}{2}}} = a^2$, so $k = 2$.

2 (a) $8^{x-3} = 4^{x+1} \Rightarrow (2^3)^{x-3} = (2^2)^{x+1}$
$\qquad \Rightarrow 2^{3(x-3)} = 2^{2(x+1)}$
$\qquad \Rightarrow 3(x - 3) = 2(x + 1)$
$\qquad \Rightarrow x = 11$.

(b) $5x^{-\frac{1}{3}} = x^{\frac{1}{6}} \Rightarrow 5x^{-\frac{1}{3}} \times x^{\frac{1}{3}} = x^{\frac{1}{6}} \times x^{\frac{1}{3}}$
$\qquad \Rightarrow 5 = x^{\frac{1}{2}}$
$\qquad \Rightarrow x = 25$.

3 (a) $x^2 - 4x + 1 = (x - 2)^2 - 3$.

(b) $(x - 2)^2 - 3 = 0$
$\qquad \Rightarrow x - 2 = \pm\sqrt{3}$
$\qquad \Rightarrow x = 2 \pm \sqrt{3}$.

(c) $(2, -3)$.

(d)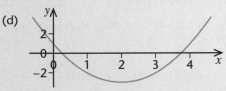

4 $f(x) = x^3 + 4x^2 + kx - 18$

(a) By the factor theorem, $f(2) = 0$
$\qquad \Rightarrow 2^3 + 4 \times 4 + 2k - 18 = 0$
$\qquad \Rightarrow 6 + 2k = 0$
$\qquad \Rightarrow k = -3$.

(b) $x^3 + 4x^2 - 3x - 18 \equiv (x - 2)(ax^2 + bx + c)$.
Equating coefficients of x^3 gives $a = 1$.
Equating the constant terms gives

$-18 = -2c \Rightarrow c = 9$.

Equating coefficients of x gives

$-3 = -2b + c = -2b + 9$
$\qquad \Rightarrow b = 6$.

So, $f(x) = (x - 2)(x^2 + 6x + 9)$
$\qquad = (x - 2)(x + 3)^2$.

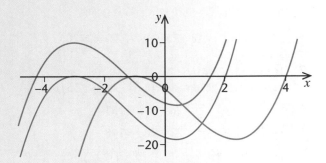

5 The co-ordinates at the points of intersection are given by the simultaneous solution of:

$y = 2x - 2$ $\qquad\qquad$ [1]

$y = x^2 - x - 6$ $\qquad\qquad$ [2]

Eliminating y gives

$\qquad\qquad x^2 - x - 6 = 2x - 2$
$\Rightarrow \qquad x^2 - 3x - 4 = 0$
$\Rightarrow \qquad (x + 1)(x - 4) = 0$
$\Rightarrow \qquad x = -1$ or $x = 4$.

When $x = -1$, $y = -4$.

When $x = 4$, $y = 6$.

The points of intersection are A$(-1, -4)$ and B$(4, 6)$.

6 (a) $4x^2 + 27x + 94 \equiv$
$\qquad\qquad$ A$(x - 1)^2 +$ B$(x - 1)(x + 4) +$ C$(x + 4)^2$.

Substituting $x = 1$ gives $125 = 25$C $\Rightarrow$ C $= 5$.
Substituting $x = -4$ gives $50 = 25$A $\Rightarrow$ A $= 2$.
Equating coefficients of x^2 gives
$4 =$ A $+$ B $+$ C
$\quad = 2 +$ B $+ 5 \Rightarrow$ B $= -3$.

(b) $3x^2 - 8x - 7 < 2x^2 - 3x - 11$
$\qquad \Rightarrow x^2 - 5x + 4 < 0$
$\qquad \Rightarrow (x - 1)(x - 4) < 0$
$\qquad \Rightarrow 1 < x < 4$.

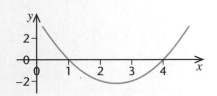

Pure 1 (continued)

7 (a) First term $a = 3$, common ratio $r = 2$.
The 8th term is $ar^7 = 3 \times 2^7 = 384$.

(b) First term $a = 25$, common difference $d = 1.2$
number of terms $n = 50$.

Using $S_n = \dfrac{n}{2}(2a + (n-1)d)$ gives

Sum $= \frac{50}{2}(2 \times 25 + 49 \times 1.2) = 2720$.

8 10th term $= a + 9d = 74$.　　　　　　　　　　[1]
Sum of first 20 terms $= 10(2a + 19d) = 1510$

$\Rightarrow$　　　　　　$2a + 19d = 151$　　　　　[2]

$2 \times$ [1] gives　　　$2a + 18d = 148$　　　　　[3]

[2] – [3] gives　　　　　$d = 3$

Subs for d in [1] gives $a + 27 = 74 \Rightarrow a = 47$.

First term $= 47$, common difference $= 3$.

9 (a) Distance travelled $=$
$2 + 2 \times (2 \times 0.75) + 2 \times (2 \times 0.75^2) + 2 \times (2 \times 0.75^3)$.
$= 8.9375 \text{ m} = 8.94 \text{ m}$ to 2 d.p.

(b) The result for part (a) may be written as
$(4 + 4 \times 0.75 + 4 \times 0.75^2 + 4 \times 0.75^3) - 2$

Using the formula for the sum of a G.P. gives the distance travelled by the nth strike as
$\dfrac{4(1 - 0.75^n)}{1 - 0.75} - 2 = 16(1 - 0.75^n) - 2$

$= 14 - 16 \times 0.75^n \text{ m}$.

(c) From part (b) the total distance travelled approaches 14 m as n increases and the term 16×0.75^n gets smaller. It can never exceed 14 m since 16×0.75^n is never negative.

10 $5 \sin x = 8 \cos x$
$\Rightarrow 5 \tan x = 8$

$\Rightarrow \tan x = 1.6$.

One solution is $x = \tan^{-1} 1.6 = 58.0°$.

Since tan repeats every 180°, the full set of solutions in the given interval is: 58.0°, 238.0°, –122.0°, –302.0°.

11 (a) $3 \sec^2 x + \tan x - 6 = 0$
$\Rightarrow 3(1 + \tan^2 x) + \tan x - 6 = 0$
$\Rightarrow 3 \tan^2 x + \tan x - 3 = 0$.

Using the quadratic formula gives
$\tan x = \dfrac{-1 \pm \sqrt{1 + 36}}{6}$

$\Rightarrow \tan x = 0.847127 \ldots$ or $\tan x = -1.18046 \ldots$.

One solution is $\tan^{-1}(0.847127 \ldots) = 40.3°$ and another is $\tan^{-1}(-1.18046 \ldots)$ so the full set of solutions in the required interval is:

40.3°, –139.7°, –49.7°, 130.3°.

12 (a)

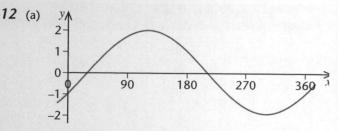

(b) $2 \sin(x - 30°) = \sqrt{3}$
$\Rightarrow \quad \sin(x - 30°) = \sqrt{3}/2$
$\sin^{-1}(\sqrt{3}/2) = 60°$.

So for the required interval, $x - 30° = 60°$ or $120°$.
This gives the solutions as 90° and 150°.

13 (a) Gradient of AB $= 6/5$.
Using $y - y_1 = m(x - x_1)$ gives
　　　$y - 1 = 6/5(x - 4)$
$\Rightarrow \quad 5y - 5 = 6x - 24$
$\Rightarrow \quad 6x - 5y - 19 = 0$.

The mid point of A and B is $(1.5, -2)$.

The gradient of the line l is $-5/6$.

Using $y - y_1 = m(x - x_1)$ again gives
　　　$y + 2 = -5/6(x - 1.5)$
$\Rightarrow \quad 6y + 12 = -5x + 7.5$
$\Rightarrow \quad 12y + 24 = -10x + 15$
$\Rightarrow \quad 10x + 12y + 9 = 0$.

At C, $x = 0 \Rightarrow y = -0.75$ so the coordinates of C are $(0, -0.75)$.

14 (a) The equation of the line m is $3x + 2y = 6$
$\Rightarrow \quad y = -3/2x + 3$
$\Rightarrow \quad$ The gradient of the line m is $-3/2$
$\Rightarrow \quad$ The gradient of the line l is $2/3$.

Using $y - y_1 = m(x - x_1)$ gives
　　　$y - 4 = 2/3(x - 5)$
$\Rightarrow \quad 3y - 12 = 2x - 10$
$\Rightarrow \quad 2x - 3y + 2 = 0$ which is the equation of line l.

At A, $y = 0 \Rightarrow x = -1$.

B lies on the line with equation $3x + 2y = 6$.

At B, $y = 0 \Rightarrow x = 2$.

So the distance AB is given by $2 - (-1) = 3$.

Pure 1 (continued)

15 (a) $y = 2x^3 - 15x^2 - 36x + 10$

$$\frac{dy}{dx} = 6x^2 - 30x - 36$$

At a stationary point

$$\frac{dy}{dx} = 0 \Rightarrow 6x^2 - 30x - 36 = 0$$

$$\Rightarrow x^2 - 5x - 6 = 0$$

$$\Rightarrow (x + 1)(x - 6) = 0$$

$$\Rightarrow x = -1 \text{ or } x = 6.$$

When $x = -1$, $y = 29$.

When $x = 6$, $y = -314$.

The stationary points are $(-1, 29)$ and $(6, -314)$.

(b) $\dfrac{d^2y}{dx^2} = 12x - 30$

When $x = -1$, $\dfrac{d^2y}{dx^2} = -42 < 0$ maximum.

When $x = 6$, $\dfrac{d^2y}{dx^2} = 42 > 0$ minimum.

The function is decreasing at all of the points between the turning points i.e. when $-1 < x < 6$.

16 (a) $x + y = 7$.

(b) At P and Q $\quad x + y = 7$ [1]

and $\quad\quad\quad\quad\quad y = (x - 1)^2 + 4$ [2]

Substituting for y in [2] gives

$$7 - x = x^2 - 2x + 5$$

$$\Rightarrow x^2 - x - 2 = 0$$

$$\Rightarrow (x - 2)(x + 1) = 0$$

$$\Rightarrow x = 2 \text{ or } x = -1.$$

$x = 2 \Rightarrow y = 5$ and $x = -1 \Rightarrow y = 8$.

P is the point $(-1, 8)$ and Q is the point $(2, 5)$.

(c) The shaded area is given by

$$\frac{3}{2}(8 + 5) - \int_{-1}^{2} x^2 - 2x + 5 \, dx$$

$$= 19.5 - \left[\frac{x^3}{3} - x^2 + 5x\right]_{-1}^{2} = 19.5 - 15 = 4.5.$$

Pure 2

1 (a) $g(x) = (x + 1)(x - 1) = x^2 - 1$.

Since $x^2 \geqslant 0$ the range of g is $x \geqslant -1$.

(b) $fg(x) = f(x^2 - 1)$

$$= 5(x^2 - 1) - 7$$

$$= 5x^2 - 12.$$

(c) $fg(x) = f(x)$

$$\Rightarrow 5x^2 - 12 = 5x - 7$$

$$\Rightarrow 5x^2 - 5x - 5 = 0$$

$$\Rightarrow x^2 - x - 1 = 0$$

$$\Rightarrow x = \frac{1 \pm \sqrt{5}}{2}.$$

2

```
              1
           1     1
        1     2     1
     1     3     3     1
  1     4     6     4     1
1     5    10    10     5     1
```

Using Pascal's triangle for the coefficients gives:

$(1 + x)^5 = 1 + 5x + 10x^2 + 10x^3 + 5x^4 + x^5$.

The term in y^6 is $10(-4y^2)^3$.

The coefficient of this term is $10 \times (-4)^3 = -640$.

3 (a) By the factor theorem $f(-2) = 0$

$$\Rightarrow (-2)^3 - 4(-2)^2 - 3(-2) + k = 0$$

$$\Rightarrow -8 - 16 + 6 + k = 0$$

$$\Rightarrow k = 18.$$

(b) $f(x) = x^3 - 4x^2 - 3x + 18$

$$\equiv (x + 2)(Ax^2 + Bx + C).$$

Comparing coefficients of x^3 gives A = 1.

Comparing the constant terms gives C = 9.

When $x = -1$

$$-1 - 4 + 3 + 18 = A - B + C$$

$$\Rightarrow 16 = 1 - B + 9 \Rightarrow B = -6.$$

$$f(x) = (x + 2)(x^2 - 6x + 9)$$

$$= (x + 2)(x - 3)^2.$$

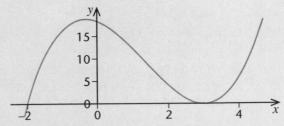

(c) From the sketch, $f(x) \geqslant 0 \Rightarrow x \geqslant -2$.

(d) By the remainder theorem, the remainder is given by $f(-1) = 16$.

Pure 2 (continued)

4 (a) $3\ln x - 2\ln y + \ln(x+1)$
$= \ln x^3 - \ln y^2 + \ln(x+1)$
$= \ln\left(\dfrac{x^3(x+1)}{y^2}\right).$

(b) $5^x = 100$
$\Rightarrow \ln 5^x = \ln 100$
$\Rightarrow x\ln 5 = \ln 100$
$\Rightarrow x = \dfrac{\ln 100}{\ln 5} = 2.861$ to 3 d.p.

(c) $\ln(2x+3) = 4$
$\Rightarrow 2x+3 = e^4$
$\Rightarrow x = \dfrac{e^4-3}{2} = 25.799$ to 3 d.p.

5 (a) LHS $\equiv \dfrac{\sin x}{\cos x} + \dfrac{\cos x}{\sin x}$
$\equiv \dfrac{\sin^2 x + \cos^2 x}{\sin x \cos x}$
$\equiv \dfrac{1}{\sin x \cos x}$
$\equiv \dfrac{1}{\sin x} \times \dfrac{1}{\cos x}$
$\equiv \sec x \operatorname{cosec} x.$

(b) $\cot^2 x + \operatorname{cosec} x = 11$
$\Rightarrow \dfrac{\cos^2 x}{\sin^2 x} + \dfrac{1}{\sin x} = 11$
$\Rightarrow \cos^2 x + \sin x = 11\sin^2 x$
$\Rightarrow 1 - \sin^2 x + \sin x = 11\sin^2 x$
$\Rightarrow 12\sin^2 x - \sin x - 1 = 0$
$\Rightarrow (4\sin x + 1)(3\sin x - 1) = 0$
$\Rightarrow \sin x = -0.25$ or $\sin x = 0.333.$
In the required interval this gives the solutions as:
$-0.253, -2.889, 0.340$ and $2.802.$

6 (a) $R\sin(x+\alpha) \equiv 15\sin x + 8\cos x$
$\Rightarrow R\cos\alpha = 15$ and $R\sin\alpha = 8$ (α is acute)
$\Rightarrow R = \sqrt{15^2 + 8^2} = 17$ giving $17\sin\alpha = 8$
$\Rightarrow \alpha = 28.07°.$
So $15\sin x + 8\cos x \equiv 17\sin(x + 28.07°).$

(b) Max value is 17.

(c) $15\sin x + 8\cos x = 12$
$\Rightarrow 17\sin(x + 28.07°) = 12.$
$\sin^{-1}\left(\tfrac{12}{17}\right) = 44.90°, 180° - 44.90° = 135.10°.$
$x + 28.07 = 44.90° \Rightarrow x = 16.8°.$
$x + 28.07 = 135.10° \Rightarrow x = 107.0°.$

7 (a) $e^{2x} - 7e^x + 6 = 0$
$\Rightarrow (e^x - 1)(e^x - 6) = 0$
$\Rightarrow e^x = 1$ or $e^x = 6$
$\Rightarrow x = 0$ or $x = \ln 6 = 1.79$ to 2 d.p.

(b) $\theta = 100e^{-0.1t}$

(i) When $t = 0$, $\theta = 100e^0 = 100$.
Temperature is 100 °C.

(ii) $100e^{-0.1t} = 50$
$\Rightarrow e^{-0.1t} = 0.5$
$\Rightarrow -0.1t = \ln(0.5)$
$\Rightarrow t = -10\ln(0.5) = 6.931\ldots$
$\Rightarrow$ time taken is 6.9 minutes to 1 d.p.

(iii) The temperature approaches zero.

(c) The temperature should approach 18 °C after a long period of time. A more refined model is $\theta = 82e^{-0.1t} + 18.$

8 (a) Area $= \displaystyle\int_1^4 3x^{\frac{2}{3}}\,dx$
$= [2x^{\frac{3}{2}}]_1^4 = 2(8 - 1)$
$= 14.$

(b) Volume $= \displaystyle\int_1^4 \pi y^2\,dx$ where $y^2 = 9x$
$= \displaystyle\int_1^4 9\pi x\,dx = 9\pi\left[\dfrac{x^2}{2}\right]_1^4$
$= 9\pi(8 - \tfrac{1}{2}) = 0\pi \times \tfrac{15}{2}$
$= \dfrac{135\pi}{2}.$

9 (a) Area $= \displaystyle\int_2^k \dfrac{5}{x}\,dx = [5\ln x]_2^k$
$= 5\ln k - 5\ln 2$
$= 5\ln\left(\dfrac{k}{2}\right).$

(b) $5\ln\left(\dfrac{k}{2}\right) = 6$
$\Rightarrow \ln\left(\dfrac{k}{2}\right) = 1.2$
$\Rightarrow k = 2e^{1.2} = 6.640$ to 3 d.p.

Pure 2 (continued)

10 $P = A\,e^{0.05t}$

(a) $\dfrac{dp}{dt} = 0.05A\,e^{0.05t} > 0$ for all values of t.

$\Rightarrow P$ is an increasing function of t.

(b) (i) Substituting $t = 10$ gives

$38 = A\,e^{0.5}$

$\Rightarrow A = 38\,e^{-0.5} = 23.04\dots$.

This represents the size of the population in millions when $t = 0$.

(ii) When $t = 20$, $P = 23.04\,e^{0.1}$

giving $P = 25.47\dots$.

So, the population is 25 million to the nearest million.

(c) $50 = 23.04\,e^{0.05t}$

$e^{0.05t} = \dfrac{50}{23.04}$

$0.05t = \ln\left(\dfrac{50}{23.04}\right)$

$t = 20\ln\left(\dfrac{50}{23.04}\right) = 15.495\dots$

The time taken is 15 years to the nearest year.

11 (a) Area sector = area triangle + area segment

$\Rightarrow \dfrac{1}{2}\theta = \dfrac{1}{2}\sin\theta + \dfrac{\pi}{5}$

$\Rightarrow \theta = \sin\theta + \dfrac{2\pi}{5}$.

This gives the iterative formula for θ as:

$\theta_{n+1} = \sin\theta_n + \dfrac{2\pi}{5}$.

(b) Taking $\theta_1 = 2$ for example, gives

$\theta_2 = 2.1659\dots$

$\theta_3 = 2.0847\dots$

$\theta_4 = 2.1274\dots$.

The sequence converges to $2.1131\dots$

So $\theta = 2.11$ to 2 d.p.

12 Area $= \displaystyle\int_2^3 \sqrt{x^2 - 3}\,dx.$

$d = \dfrac{3 - 2}{5} = 0.2$ Taking $f(x) = \sqrt{x^2 - 3}$ gives

$A = 0.1(f(2) + 2(f(2.2) + f(2.4) + f(2.6) + f(2.8)) + f(3))$

$A = 0.1(1 + 2(1.3565 + 1.6613 + 1.9591 + 2.2) + 2.4495)$

$A = 1.78$ to 2 d.p.

13 (a) (i) $y = (4x - 3)^9$

$\Rightarrow \dfrac{dy}{dx} = 9(4x - 3)^8 \times 4$

$\Rightarrow \dfrac{dy}{dx} = 36(4x - 3)^8.$

(ii) $y = \ln(5 - 2x)$

$\dfrac{dy}{dx} = \dfrac{1}{5 - 2x} \times (-2)$

$= \dfrac{-2}{5 - 2x}$ $\left(\text{you can write this as } \dfrac{2}{2x - 5}\right).$

(b) (i) $\displaystyle\int (4x - 3)\,dx = \dfrac{1}{36}(4x - 3)^9 + C.$

(ii) $\displaystyle\int_3^4 \dfrac{1}{5 - 2x}\,dx = -\dfrac{1}{2}\int_3^4 \dfrac{-2}{5 - 2x}\,dx$

$= -\tfrac{1}{2}\,[\ln|5 - 2x|]_3^4$

$= -\tfrac{1}{2}\,(\ln|-3| - \ln|-1|)$

$= -\tfrac{1}{2}\ln 3.$

14 $\dfrac{dv}{dt} = 200$, $v = \dfrac{4}{3}\pi r^3.$

$\dfrac{dv}{dt} = \dfrac{dv}{dr} \times \dfrac{dr}{dt}$

$\Rightarrow 200 = 4\pi r^2 \times \dfrac{dr}{dt}$

$\Rightarrow \dfrac{dr}{dt} = \dfrac{200}{4\pi r^2} = \dfrac{200}{4\pi \times 25}$

$\Rightarrow \dfrac{dr}{dt} = \dfrac{2}{\pi}.$

The radius increases at the rate of $\dfrac{2}{\pi}$ cm s^{-1}.

Mechanics 1

1 (a)

$$u = 10$$
$$a = -4$$

Using $v = u + at$ $v = 10 - 4 \times 5 = -10$.

The speed of the particle when $t = 5$ is 10 m s^{-1}.

(b)

$$v = 0 \Rightarrow 10 - 4t = 0$$
$$\Rightarrow t = 2.5.$$

Using $\qquad\qquad s = ut + \frac{1}{2}at^2$

when $t = 2.5$ $\qquad s = 10 \times 2.5 - 2 \times 6.25$

$$s = 12.5.$$

when $t = 4$ $\qquad s = 10 \times 4 - 2 \times 16$

$$s = 8.$$

The total distance travelled between $t = 0$ and $t = 4$ is 12.5 m + 4.5 m = 17 m.

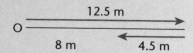

2 (a) The acceleration is positive between $t = 0$ and $t = 25$ so the greatest velocity occurs when $t = 25$.

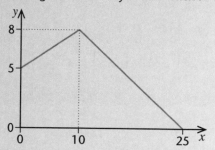

The area under the graph represents the increase in velocity between $t = 0$ and $t = 25$.

Area $= \frac{1}{2} \times 10(5 + 8) + \frac{1}{2} \times 15 \times 8 = 125$.

The increase in velocity is 125 m s^{-1}.

The initial velocity is 20 m s^{-1}.

The maximum value of the velocity is 145 m s^{-1}.

3 (a) The acceleration is constant so the constant acceleration formulae may be used.

$$\mathbf{u} = 2\mathbf{i} + \mathbf{j}$$
$$\mathbf{a} = \mathbf{i} + 3\mathbf{k}$$
$$t = 4.$$

Using $\mathbf{s} = \mathbf{u}t + \frac{1}{2}\mathbf{a}t^2$

$$\mathbf{s} = (2\mathbf{i} + \mathbf{j})4 + (\mathbf{i} + 3\mathbf{k})8$$
$$\mathbf{s} = 16\mathbf{i} + 4\mathbf{j} + 24\mathbf{k}.$$

This is the displacement from P to Q

so $\qquad \overrightarrow{PQ} = 16\mathbf{i} + 4\mathbf{j} + 24\mathbf{k}$.

(b) $\overrightarrow{OQ} = \overrightarrow{OP} + \overrightarrow{PQ}$

$$= -5\mathbf{i} + 11\mathbf{j} + \mathbf{k} + 16\mathbf{i} + 4\mathbf{j} + 24\mathbf{k}$$
$$= 11\mathbf{i} + 15\mathbf{j} + 25\mathbf{k}.$$

(c) Average speed $= \dfrac{\text{distance}}{\text{time}} = \dfrac{|\overrightarrow{PQ}|}{5}$ m s^{-1}

$$|\overrightarrow{PQ}| = \sqrt{16^2 + 4^2 + 24^2} = 29.12.$$

Average speed $= 5.82$ m s^{-1} to 2 d.p.

4

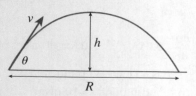

vertically

$$u = v \sin \theta$$
$$a = -g$$

At max height $\quad v = 0$.

Using $\qquad\qquad v = u + at$

$$0 = v \sin \theta - gt$$

$\Rightarrow \qquad\qquad t = \dfrac{v \sin \theta}{g}.$

Using $\qquad\qquad s = ut + \frac{1}{2}at^2$

$$h = v \sin \theta . \frac{v \sin \theta}{g} - \frac{1}{2}g\left(\frac{v \sin \theta}{g}\right)^2$$

$$h = \frac{v^2 \sin^2 \theta}{2g}.$$

The motion is symmetrical, so total time of flight is 2 × time taken to reach the greatest height.

Time of flight $= \dfrac{2v \sin \theta}{g}$.

Horizontal component of velocity $= v \cos \theta$.

Range $= v \cos \theta \times \dfrac{2v \sin \theta}{g} = \dfrac{v^2(2 \sin \theta \cos \theta)}{g}$

$$= \frac{v^2 \sin 2\theta}{g}.$$

5

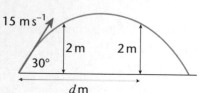

Vertically

$$u = 15 \sin 20°$$
$$a = -9.8$$
$$s = 2.$$

Using $\qquad s = ut + \frac{1}{2}at^2$

$$2 = 15 \sin 30° t - 4.9t^2$$

$\Rightarrow 4.9t^2 - 15 \sin 30° t + 2 = 0$

$\Rightarrow t = \dfrac{15 \sin 30° \pm \sqrt{(15 \sin 30°)^2 - 8 \times 4.9}}{9.8}$

$\Rightarrow t = 1.186\dots$ or $t = 0.3437\dots$

(*there are two times when the ball has height 2 m*)

$$d = 1.186\dots \times 15 \cos 30°$$

The maximum distance = 15.4 m to 1 d.p.

Mechanics 1

6

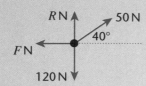

(a) Resolving vertically
$$R + 50 \sin 40° - 120 = 0$$
$$\Rightarrow R = 120 - 50 \sin 40°$$
$$\Rightarrow R = 87.86 \ldots.$$

The normal reaction is 87.9 N to 1 d.p.

Resolving horizontally
$$F - 50 \cos 40° = 0$$
$$\Rightarrow F = 38.30 \ldots.$$

For limiting equilibrium
$$F = \mu R$$

giving
$$38.30\ldots + \mu \times 87.86\ldots$$
$$\mu = 0.4359 \ldots.$$

The coefficient of friction is 0.436 to 3 d.p.

7

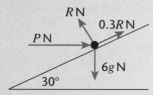

The equations are simpler to deal with by resolving vertically and horizontally, in this case, instead of parallel and normal to the plane.

The minimum value of P corresponds to the case where friction is limiting, as shown in the diagram.

Resolving vertically
$$R \cos 30° + 0.3 R \sin 30° = 6 \times 9.8$$

$$R = \frac{6 \times 9.8}{\cos 30° + 0.3 \sin 30°}$$

$$R = 57.87 \cdots.$$

Resolving horizontally
$$P + 0.3R \cos 30° - R \sin 30° = 0$$

$$P = R \sin 30° - 0.3R \cos 30°$$

$$P = 13.90 \ldots.$$

The minimum value of P is 13.9 to 1 d.p.

8

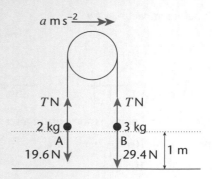

(a) The equations of motion for the particles are:
For A $\qquad T - 19.6 = 2a$
For B $\qquad 29.4 - T = 3a$.
Adding gives $9.8 = 5a \Rightarrow a = 1.96$.
The acceleration of the system is 1.96 m s^{-2}

(b) $T - 19.6 = 3.92 \Rightarrow T = 23.52$.
The tension in the string is 23.52 N.

(c) For B $\quad u = 0$
$\qquad a = 1.96$,
$\qquad s = 1$.
Using $\quad s = ut + 1/2at^2$
$\qquad 1 = 0 + 0.98t^2 \; t > 0$
$\qquad \Rightarrow t = 1.01 \ldots.$
Particle B hits the ground after 1.01 s to 2 d.p.

(d) using $\quad v = u + at$
$\qquad v = 0 + 1.96 \times 1.01\ldots = 1.979 \ldots.$
Particle B hits the ground with speed 1.98 m s^{-1} to 2 d.p.

(e) Particle A rises to height 2 m then the string becomes slack and it behaves as a projectile.
Using $\qquad v^2 = u^2 + 2as$
at max height $\qquad 0 = 1.979^2 - 19.6s$
$\qquad\qquad \Rightarrow s = 0.2$.
The max height reached by particle A is 2.2 m.

(f) In the extreme situation where the pulley does not move, each particle would be held in equilibrium and so the tension on each side would match the weight of the corresponding particle.

In the given situation, where the pulley is not light or frictionless, it cannot be assumed that the tension on each side of the pulley is the same. This introduces an extra unknown value into the equations. For example, the tension on one side might be labelled T_1 and on the other T_2.

Mechanics 1

9   Before impact $5u$

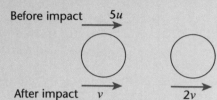

After impact v $2v$

Momentum before = momentum after

so $10mu = 2mv + 2mv$

$\Rightarrow 10u = 4v$

$\Rightarrow v = 2.5u$

(b)   Impulse = change in momentum

$= 5mu - 0$

$= 5mu.$

10

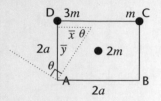

(a) $\bar{x} = \dfrac{3m \times 0 + 2ma + 2ma}{6m} = \dfrac{4ma}{6m}$

$\bar{x} = \dfrac{2a}{3}$

(b) $\bar{y} = \dfrac{2ma + 6ma + 2ma}{6m} = \dfrac{10ma}{6m}$

$\bar{y} = \dfrac{5a}{3}.$

(c) $\tan \theta = \dfrac{\bar{y}}{\bar{x}} = \dfrac{5}{2} \Rightarrow \theta = 68.2°.$

AD makes an angle of 68.2° with the horizontal.

Statistics 1

1   (a)   Care is needed because the letter P is repeated. The key is to deal with this systematically. Any selection must contain 0, 1 or 2 Ps.

No. of Ps in selection	No. of selections
0	1 ($^4C_4$)
1	4 ($^4C_3$)
2	6 ($^4C_2$)
Total	11

(b)

No. of Ps in arrangement	No. of arrangements
0	$1 \times 4! = 24$
1	$4 \times 4! = 96$
2	$6 \times = \dfrac{4!}{2} = 72$
Total	192

2   (a) $\Sigma p(x) = 1$

$\Rightarrow \dfrac{1}{k} + \dfrac{4}{k} + \dfrac{9}{k} = 1$

$\Rightarrow \dfrac{14}{k} = 1$

$k = 14.$

(b)   Mean $= E(X) = \Sigma x p(x)$

$= \dfrac{1}{14} + 4 \times \dfrac{4}{14} + 3 \times \dfrac{9}{14} = \dfrac{36}{14}$

$= \dfrac{18}{7}.$

$\text{Var}(X) = E(X^2) - (E(X))^2$

$= x^2 p(x) - \left(\dfrac{18}{7}\right)^2$

$= \dfrac{1}{14} + 4 \times \dfrac{4}{14} + 9 \times \dfrac{9}{14} - \left(\dfrac{18}{7}\right)^2$

$= \dfrac{19}{49}.$

(c)   Mean of $3X - 5$ is $3 \times \dfrac{18}{7} - 5 = \dfrac{19}{7}.$

Variance of $3X - 5$ is $9 \times \dfrac{19}{49} = \dfrac{171}{49}.$

Statistics 1

3 $X \sim B(20, 0.4)$ so $n = 20$ and $p = 0.4$.

(a) Mean of $X = np = 20 \times 0.4 = 8$

Variance of $X = np(1 - p) = 20 \times 0.4 \times 0.6$
$$= 4.8.$$

(b) $P(2 \leqslant X \leqslant 4) = P(X = 2) + P(X = 3) + P(X = 4)$.

$$= \binom{20}{2}(0.4)^2(0.6)^{18} + \binom{20}{3}(0.4)^3(0.6)^{17}$$

$$+ \binom{20}{2}(0.4)^4(0.6)^{16}$$

$$= 0.00309 + 0.01235 + 0.03499 = 0.05043$$

$$= 0.050 \text{ to 3 d.p.}$$

(c) $P(X \geqslant 5) = 1 - P(X \leqslant 4)$

$$= 1 - (P(X = 0) + P(X = 1) + P(2 \leqslant X \leqslant 4))$$

$$= 1 - (0.6^{20} + 20 \times 0.4 \times 0.6^{19} + 0.05043)$$

$$= 0.949 \text{ to 3 d.p.}$$

4 (a) $P(A \cup B) = P(A) + P(B) - P(A \cap B)$
$0.72 = 0.3 + 0.6 - P(A \cap B)$
$\Rightarrow P(A \cap B) = 0.18$.
As $P(A \cap B) \neq 0$,
A and B are not mutually exclusive.

(b) $P(A) \times P(B) = 0.3 \times 0.6 = 0.18$.
As $P(A \cap B) = P(A) \times P(B)$,
A and B are independent.

(c) $P(A \cup B)' = 1 - P(A \cup B) = 0.28$.

(d) $P(A' \cup B') = P(A') + P(B') - P(A' \cap B')$

$$= 0.7 + 0.4 - 0.7 \times 0.4$$

$$= 0.82.$$

5 (a) (i) $P(RRR) = \dfrac{26}{52} \times \dfrac{25}{51} \times \dfrac{24}{50} = \dfrac{2}{17}$

(ii) $P(\text{same colour}) = P(RRR) + P(BBB)$

$$= \frac{2}{17} + \frac{2}{17} = \frac{4}{17}.$$

(iii) $P(\text{at least one red}) = 1 - P(BBB)$

$$= 1 - \frac{2}{17} = \frac{15}{17}.$$

(b) Using A to represent getting an Ace and D to represent getting a Diamond

$$P(A|D') = \frac{P(A \cap D')}{P(D')} = \frac{\frac{3}{52}}{\frac{3}{4}} = \frac{3}{52} \times \frac{4}{3}$$

$$= \frac{1}{13}.$$

6 (a) (i) $P(X = 0) = e^{-0.8} = 0.449$ to 3 d.p

(ii) $P(X = 0) + P(X = 1) + P(X = 2)$
$$= 0.4493 + 0.3595 + 0.1438$$
$$= 0.9526$$
$$= 0.953 \text{ to 3 d.p.}$$

(b) for 4 weeks, mean $= 0.8 \times 4 = 3.2$
$P(X = 0) = e^{-3.2} = 0.041$ to 3 d.p

7 Using X to represent the lifetime of a battery in hours:

(a) $P(X > 8.5) = P\left(Z > \dfrac{8.5 - 8}{\frac{25}{60}}\right)$

$$= P(Z > 1.2)$$

$$= 1 - \Phi(1.2)$$

$$= 1 - 0.8849 \dots$$

$$= 0.1151 \dots .$$

So the probability that a battery selected at random will last more than $8\frac{1}{2}$ hours is 0.115 to 3 d.p.

(b) $P(X < 7) = P\left(Z < \dfrac{7 - 8}{\frac{25}{60}}\right)$

$$= P(Z < -2.4)$$

$$= \Phi(-2.4)$$

$$= 1 - \Phi(2.4)$$

$$= 1 - 0.9918$$

$$= 0.0082.$$

The percentage of batteries that will last for less than 7 hours is 0.82%.

8 (a)

Teacher	Teacher	Rankings			
A	B	A	B	d	d^2
39	46	5	5	0	0
72	80	1	1	0	0
51	53	3	4	−1	1
49	54	4	3	1	1
68	73	2	2	0	0
					$d^2 = 2$

$$r_s = 1 - \frac{6 \times 2}{5^3 - 5} = 0.9.$$

(b) No. A high correlation produced in this way means that the markers agree closely in terms of the *relative* merit of the pieces of work but not necessarily in terms of the actual marks. For example, one marker may consistently award a higher mark than the other.

Statistics 1

9 (a) Taking X to represent the number of packets in a box that contain a lucky pendant:

(i) $P(X \geqslant 6) = 1 - P(X \leqslant 5)$

Using tables of cumulative binomial probabilities with $n = 20$ and $p = 0.25$.

$P(X \geqslant 6) = 1 - 0.6172 = 0.3828$.

(ii) $P(X = 6) = P(X \leqslant 6) - P(X \leqslant 5)$

$= 0.7858 - 0.6172$

$= 0.1686$.

(b) Using Y to represent the number of boxes that have at least six packets containing a lucky pendant:

$P(Y = 2) = \binom{5}{2}(0.1686)^2(1 - 0.1686)^3$

$= 0.163359\ldots$

$= 0.163$ to 3 d.p.

10 (a) $r = \dfrac{S_{xy}}{\sqrt{S_{xx}S_{yy}}}$

$= \dfrac{28785 - \dfrac{567 \times 484}{10}}{\sqrt{\left\{\left(34216 - \dfrac{576^2}{10}\right)\left(24450 - \dfrac{484^2}{10}\right)\right\}}}$

$= 0.879$ to 3 d.p.

(b) $d = \dfrac{S_{xy}}{S_{yy}} = \dfrac{28785 - \dfrac{576 \times 484}{10}}{24450 - \dfrac{484^2}{10}}$

$= 0.885$ to 3 d.p.

$c = \dfrac{576}{10} - 0.8850 \times \dfrac{484}{10} = 14.766$

The equation is $x = 14.8 + 0.885y$.

(c) (i) $14.8 + 0.885 \times 42 = 51.97$.

This gives an estimate of 52 marks.

(ii) A physics score of 80 is outside the interval of data for the explanatory variable so the result would be very unreliable.

Decision 1

1 (a) Total memory to transfer = 4520 Kb = 4.52 Mb

$\dfrac{4.52}{1.4} = 3.228\ldots$

Three disk do not have sufficient capacity to hold all of the files. A minimum of four disks is required.

(b)
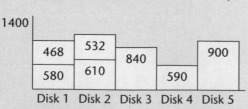

(c) Writing the memory sizes in order gives

900, 840, 610, 590, 580, 532, 468.

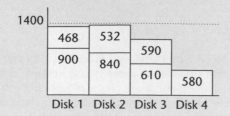

2 Applying the matrix form of Prim's algorithm gives:

	1	7	5	6	3	2	4
	A	B	C	D	E	F	G
A	–	50	40	61	–	38	–
B	50	–	–	–	–	(47)	54
C	(40)	–	–	35	–	–	–
D	61	–	(35)	–	42	–	–
E	–	–	–	42	–	(25)	10
F	(38)	47	–	–	25	–	63
G	–	54	–	–	(10)	63	–

The minimum cost is found by adding the circled figures and multiplying by £100.

Minimum cost = £19 500.

Decision 1 *(continued)*

3

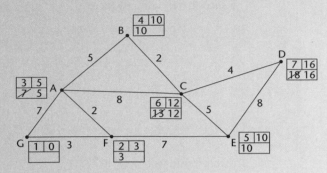
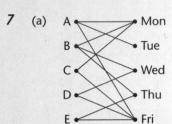
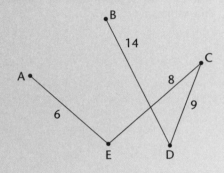
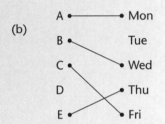
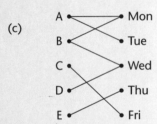

(a) The shortest distance from G to D is 16.

(b) The route is G, F, A, B, C, D.

4

B

14

A

C

8

9

6

E   D

(a) The order in which the edges were connected is:

AE, EC, CD, AB (not AD or AC to avoid cycles).

The length of the minimum spanning tree is 37.

(b) Using the short-cut from B to A gives an upper bound for the TSP as 37 + 12 = 49.

5

Vertices				
Graph (i) A	B	C	D	E
Order 4	2	4	3	3

Vertices					
Graph (ii) A	B	C	D	E	F
Order 4	2	4	4	4	2

Vertices			
Graph (iii) A	C	D	E
Order 3	3	3	3

(b) Graph (i) is Semi-Eulerian (2 odd vertices).

Graph (ii) is Eulerian (all vertices are even).

Graph (iii) is neither (more than two odd vertices).

6 (a) The objective function is $4a + 9b + 12c$. The formulation of the task as a linear programming problem is:

Maximise: $P = 4a + 9b + 12c$

Subject to: $6a + 15b + 20c \leqslant 100$

$$a + b + c \leqslant 10$$

and $a \geqslant 0, b \geqslant 0, c \geqslant 0$.

(c) In the solution a, b and c must be integers.

7 (a)

A   Mon

B   Tue

C   Wed

D   Thu

E   Fri

(b)

A   Mon

B   Tue

C   Wed

D   Thu

E   Fri

(c)

A   Mon

B   Tue

C   Wed

D   Thu

E   Fri

The alternating path is:

D – Wed – B – Mon – A – Tue.

The new matching is:

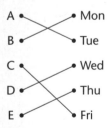

A   Mon

B   Tue

C   Wed

D   Thu

E   Fri

Decision 1 *(continued)*

8 (a) The initial tableau is:

P	x	y	z	s	t	val
1	−1	−2	−3	0	0	0
0	2	6	1	1	0	10
0	1	4	5	0	1	16

(b) The first iteration produces:

P	x	y	z	s	t	val
1	$-\frac{2}{5}$	$\frac{2}{5}$	0	0	$\frac{3}{5}$	$9\frac{3}{5}$
0	$\frac{9}{5}$	$\frac{26}{5}$	0	1	$-\frac{1}{5}$	$6\frac{4}{5}$
0	$\frac{1}{5}$	$\frac{4}{5}$	1	0	$\frac{1}{5}$	$3\frac{1}{5}$

The second iteration produces:

P	x	y	z	s	t	val
1	0	$\frac{14}{9}$	0	$\frac{2}{9}$	$\frac{5}{9}$	$11\frac{1}{9}$
0	1	$\frac{26}{9}$	0	$\frac{5}{9}$	$-\frac{1}{9}$	$3\frac{7}{9}$
0	0	$\frac{2}{9}$	1	$-\frac{1}{9}$	$\frac{6}{25}$	$2\frac{4}{9}$

The maximum value of P is $11\frac{1}{9}$.

This occurs when $x = 3\frac{7}{9}$, $y = 0$, $z = 2\frac{4}{9}$.

(c) This represents the optimum solution because there are no negative values in the objective row of the tableau.

Index

acceleration 72, 75–8, 84–5
activity network 120
addition rule 96
algebra 19–28, 48–55
algorithms 111–13, 115–19, 123, 125
angle formulae 57–8
area under a curve 42, 66
arithmetic progression 33–4
average 94

binary search 113
binomial expansion 51–2
binomial probability distribution 100
bipartite graph 125
bounds 117
box and whisker plot 95
bubble sort 112

calculus 37, 41
capacity 118–19
centre of mass 83
chain rule 61
collision 86–7
combinations 98
component form vectors 73–4, 79
compound angle formulae 57–8
conditional probability 97
connected particles 85
conservation of momentum 86
constant acceleration 75, 77
continuous random variable 102
coordinate geometry 30–2
correlation 104
correlation coefficient 104
cosecant 56, 58
cosine 35–6, 56–9
cotangent 56
critical path 120–1
cubic curve 24

cumulative distribution 99
curve sketching 39
cut 118

data 94–6
decimal search 64
dependency table 120
derivative 38–40
differentiation 38–40, 41, 60–1
Dijkstra's algorithm 116
direction 72, 74, 77
discrete random variables 99–101
dispersion 95
displacement 72, 75, 76
distance 72, 76
domain 20, 48, 57
double angle formulae 58

earliest event time 120
equation of motion 85
equilibrium 80–1, 82
Euler 114, 118
expected value 99–100
exponential equations 55
exponential logarithm 54, 60, 62
extrapolation 105

factor theorem 27, 53
factorising 21, 27–8
feasible region 122
first-fit algorithm 113
flow augmentation 119
flow diagram 111
flows 118–19
force 72, 79–80, 84–7
force diagram 79
frequency distribution 94
friction 79, 81
friction coefficient 81
full-bin algorithm 113
function 20, 48–51

geometric distribution 101
geometric progression 33, 34
geometry 30–2

gradient 30, 31, 38, 54
graphical terms 114
gravity 75, 78, 79
grouped data 94

half angle formulae 58

identity 26, 36, 57–9
image 20
impulse 86
inclined plane 79–80
independent events 97
indices 19
inequalities 28
integral 42
integration 41–2, 54, 62–3, 66
integration constant 41
intercept 30, 31
inter-quartile range 95
inverse of a function 48–50
inverse trigonometric functions 57
iteration 65, 124

kinematics 75–8
Kruskal's algorithm 115

lamina 83
latest event time 120
least squares regression 105
light inextensible string 85
light rigid bodies 83
log laws 55, 60
logarithm 54–5, 60

magnitude 72, 74, 80
map 20
mass 72, 79, 81, 83–7
matching improvement algorithm 125
maximum 39–40
maximum flow-minimum cut theorem 119
mean 94, 99, 102–3
median 94
mid-point 32
minimum 39–40

mode 94
modulus function 25, 50–1
moments 82
momentum 72, 84, 86
multiplication rule 97
mutually exclusive events 96–7

natural logarithm 54, 60
nearest neighbour algorithm 117
Newton's laws of motion 84–5
normal 39
normal distribution 102–3
normal reaction 79

parameters 100
Pascal's triangle 51–2
periodic function 35, 56
permutations 97–8
pivot 112
pivotal column & row 123–4
point of inflexion 39
Poisson distribution 101
polynomials 27, 52–3
position vector 73, 77
precedence table 120
Prim's algorithm 115
probability 96–101
probability density function 102
probability distribution 99
programming 122–4
projectiles 78
Pythagoras' theorem 35
Pythagorean identities 57–9

quadratic graphs 23
quadratics 21–3, 26–8, 59
quick sort 112
quotient 52–3

radians 37
random variable 99–100, 102

Index

range 20, 48, 95
reaction 79
reciprocal function 62
recurrence relation 33
recursive definition 33
reflection 50
regression 105
remainder 52–3
remainder theorem 53
residuals 105
resolving a vector 73–4
resolving forces 79–80
resultant force 80, 82, 84
resultant vector 72
route inspection problem 118

scalar multiplication 73
scalar quantity 72

scatter diagram 104
scheduling 121
secant 56, 58
second derivative 39–40
sequences 33–4
series 33–4
sets 96
Simplex algorithm 123
simultaneous equations 26, 30–1, 85, 103
sine 35–7, 56–9
skew 95
speed 72, 76–8, 86–7
square root 19
standard deviation 95, 100, 102–3
standard normal variable 102
stationary points 39–40
statistical diagrams 95

statistics 94–105
stem & leaf diagram 95
straight line equation 30–2
straight line motion 75–6
substitution 26–7, 30–1
surds 19
symmetry 24, 35, 56

tableau 123–4
tangent 35–6, 56–8
tangent to a curve 39
tension 79, 85
time 72, 75, 76, 86–7
total float 121
transformations 25, 36
trapezium rule 66
travelling salesman problem 117

tree diagram 97
trigonometry 35–7, 56–9
turning effect 82

uniform bodies 83
unit vectors 73

variable acceleration 76–7
variance 95, 100, 103
vector diagrams 79
vectors 72–4, 77, 86
velocity 72, 75–8, 85, 86–7
venn diagrams 96
vertex 23, 114–18, 120, 123, 125
volume of revolution 63

weight 78–80, 81